Even Mathematicians and Physicists Make Mistakes

By

Dennis Morris

© Dennis Morris

Published by: Abane & Right

56 Coach Road

Brotton

Saltburn

TS12 2RP

United Kingdom

01287 678918

dennis355@btinternet.com

August 2016

Contents

Introduction 1
    Physicists are humble: 2
    Mathematicians are not so humble: 2
    About this book: 4
The Nature of Mathematical Proof 6
    The social side of mathematical proof: 6
    Theorema egregium: 6
    Fermat's last theorem: 7
    Complicated proofs: 8
    The Classification theorem: 9
    Summary of social side of mathematical proof: 10
    Undeclared assumptions: 11
    Undeclared assumptions of Riemann geometry: 11
    Another undeclared assumption: 12
    Computer proofs: 13
    The Goldbach conjecture: 14
    Proof is something more than proof: 15
    Summary: 16
Two Ways to Construct Mathematics 17
    The axiomatic system: 17
    Deduction from the number one: 18
    Two ways of building mathematics: 19
Introduction to the Ensuing Chapters 21
The Algebraic Extension Error 22
    The wrong road: 23
    Consequences of the error: 24

Even Mathematicians and Physicists make Mistakes

More history: 26
The error explained: 26
The finite group road: 27
Interim summary: 28
Notation: 29
The mathematics: 31
The revolution: 33
Addendum: 34
The Hyperbolic Complex Numbers Error 36
Special relativity: 38
An error of forgetfulness: 38
Careless Errors 39
Neutrino mass: 40
The neutrino field is massless: 40
The neutrino field squared is massive: 41
Massive, massless, which is it?: 41
Special relativity is 2-dimensional: 42
The teaching of special relativity: 44
The petty magnetic dipole error: 45
Conservation of energy properly presented: 45
Wave-particle duality: 46
Summary: 48
The All Rotations are 2-dimensional Error 49
Higher dimensional rotations: 49
Higher dimensional rotations again: 50
Summary: 52
Θersted and the Breakdown of Parity Error 54

Contents

The discovery of electromagnetism: 54

The magnetic field is left-handed: 56

Interim summary: 58

Missed special relativity: 59

The nature of our space-time: 59

The Rotation about an Axis Error 61

For physicists: 63

No rotation in the complex plane: 63

The uniqueness of our 4-dimensional space-time: 64

Genesis of the spin axis error: 65

The Complex Numbers Differentiation Error 66

Differentiation within the complex numbers: 66

Integration: 67

Unsatisfactory differentiation: 67

Differentiation within the complex numbers, $\mathbb{C}$ - a different view: 68

Matrix differentiation: 69

A complex potential: 71

Differential operators: 71

The Cauchy-Riemann differential is included: 73

Differentiation in 2-dimensional space-time: 74

Differentiation of a 3-dimensional division algebra: 75

All differentiation: 76

Double differentiation: 76

Summary: 76

How did mathematicians miss this?: 77

Summary: 78

Non-Commutative Differentiation 79

    The differential: 79

    Perhaps we are being too harsh: 81

The Empty Space Error 82

    Distance functions do not hold angles: 84

    Spinor spaces: 84

    Our 4-dimensional space-time: 86

    The quaternion emergent expectation space: 86

    The empty space error restated: 87

The Error of Hilbert Spaces 89

    Hilbert space definition: 90

    Multiplying vectors together: 91

    The Hilbert space inner products: 91

    The Hilbert space error: 92

    Complex vectors: 93

    Interim summary: 94

    How did this error occur?: 94

    Return to reality: 95

    Hilbert spaces are useful: 96

    Summary: 96

The Error of Pauli's Spinors 97

    Confusion upon confusion: 98

    As a quaternion: 99

    A notational error: 99

    To err with notation is human: 100

The Dirac Equation Error 101

    But the Dirac equation is wrong: 102

    The Schrödinger equation: 102

Contents

- Let us take pause for thought: 103
- Onto relativistic quantum theory: 104
- The Dirac equation: 105
- Four roots: 106
- The Dirac error: 106
- Clifford algebras: 106
- Dirac's first error: 108
- Dirac's second error: 112
- Dirac's third error: 113
- Majorana particles: 114
- Dirac's great triumph: 114
- Dirac's great oversight: 115
- Genesis of the error: 115
- Appendix – List of papers of interest: 117

The Scattering Amplitude Error 119
- A promulgated error: 120
- Origin of this error: 120

The Lie Group Error 121
- Rotation: 121
- Lorentz invariant physics: 122
- Rotation in a spinor space: 123
- Spinor rotation as seen from within our space-time: 124
- Lie groups: 125
- Lie groups and infinitesimal transformations: 125
- The error of Lie algebras: 126
- The first misapprehension: 126
- The second misapprehension: 127

- Why do we have Lie groups in physics?: 128
- Postmethian comments: 129
- Concluding Remarks 130
  - Notation: 131
  - Better to err than to do nothing: 131
  - Don't believe the stark raving obvious: 132
  - We walk a rugged road: 132
  - The biggest error: 132
- Other Books by the Same Author 133
- Index 137

Introduction

Chapter 1

Introduction

In any area of human endeavour, there are giants who stand head and shoulders above their colleagues. In mathematics, names like Euclid (325 BC - 265 BC), Johann Karl Friedrich Gauss (1777-1855), Augustin-Louis Cauchy (1789-1857), Bernhard Riemann (1826-1866), David Hilbert (1862-1943), and Bertrand Russell (1872-1970), stand out from the crowd. In physics, names like Aristotle (384 BC – 322 BC), Isaac Newton (1643-1727), Erwin Schrödinger (1887-1961), Albert Einstein (1879-1955), Wolfgang Pauli (1900-1958), Paul Adrien Maurice Dirac (1902-1984), and Richard Feynman (1918-1988) stand out from the crowd. There are of course many other outstanding individuals who we have not named.

Such names are often so venerated by their contemporaries, and more so by the generations which follow such names, that the contributions these great individuals make to their fields of endeavour are by many considered too sacrosanct to possibly be in error. Often for centuries after these great people have lived, no-one will seriously dare to challenge the contributions made by these people. Yet, history has already shown that some of these paragons of intellect did make mistakes.

This failure to challenge is understandable. We use Dirac as an example.

1) Firstly, if Dirac made a mistake, it must have been a subtle one. If Dirac's mistake was so subtle that Dirac and his contemporaries did not see it, then it is unlikely that the lesser intellect of a young undergraduate will see that mistake.
2) Secondly, what future is there for a young undergraduate who has the gall to challenge Dirac? So she does not challenge Dirac, and who would listen if she did. By the time the young undergraduate has become an established academic, much of

her own work will be built upon the work of Dirac, and there are social and career costs involved in pointing out errors which others have not seen and upon which others have built their own reputations. It takes a person of considerable arrogance to challenge an established great individual; most individuals lack either the self-belief or the arrogance to voice such challenge.
3) Thirdly, the great individuals are considered great because the contributions they made work and work extremely well. The equation for which Dirac is famous works as perfectly as we can measure; how can it be in error?

Yet, centuries after they lived, we now know that even the towering intellects of Euclid, Aristotle, and Isaac Newton made mistakes. Is it possible than more recent towering intellects like David Hilbert or Richard Feynman have also made mistakes?

Physicists are humble:
Undoubtedly, all of the great physicists, if they are or were alive today, would be the first to declare they believe themselves to have made mistakes, but they would add that they have no idea of the nature of these mistakes. "If we had known what the mistakes were, we would not have made them", they might declare. Alternatively, they might say, "Yes, I know we swept a minus sign under the carpet, but it worked, and we had no idea how to make it work without the sweeping brush and the carpet". They might add, "One does not need to look far into the history of physics to find great physicists who got it wrong", and, with this in mind, the great physicists are most cautious about the certainty of their contributions.

Mathematicians are not so humble:
Mathematicians are different; they work on proof. Mathematicians believe that everything ever done by any mathematician is correct because it has been proven to be correct. Physicists, even great ones, are judged by observation of reality. There is no way to be certain that tomorrow will not see the report of an observation which clearly shows

Introduction

there is a mistake within our physical theories. Mathematicians are judged by, or believe themselves to be judged by, pure logic. If the mathematics is done properly, then finding a mistake would be to throw logic out of the window. Mathematicians are the holders of absolute truth. It might well be that reality does not fit with mathematics, but that is a problem for reality. Why should the universe be commensurate with mathematics?

Mathematicians have good reason to be confident in the correctness of their proven mathematics. Let us consider a proof of the irrationality of the square root of two. Working in base ten, we calculate the square of numbers with various final, 10^0, digits. For example, a number that ends in a 0 will have a square that ends in a 0 and a number that ends in a 2 will have a square that ends in a 4. We get:

$$
\begin{aligned}
&(....0)^2 = ...0 \quad (....1)^2 = ...1 \quad (....2)^2 = ...4 \\
&(....3)^2 = ...9 \quad (....4)^2 = ...6 \quad (....5)^2 = ...5 \\
&(....6)^2 = ...6 \quad (....7)^2 = ...9 \quad (....8)^2 = ...4 \\
&(....9)^2 = ...1
\end{aligned}
\quad (1.1)
$$

We see that there are, in base ten, no square numbers which have a $\{2,3,7,8\}$ as a final digit. Thus we have proven that any number, written in base ten, ending in $\{2,3,7,8\}$ does not have a square root with a final digit – does not have a rational number for a square root. How can this proof be in error? Your author suspects this proof cannot be in error. With such proofs, mathematicians are justified in being confident that they have made no mistakes.

Things are not that simple, and the above proof, (1.1), needs some qualification. Above, (1.1), we have proven that numbers, in base ten, ending with $\{2,3,7,8\}$ do not have rational square roots. We have not proven that numbers, in base ten, ending with $\{2,3,7,8\}$ have irrational square roots. It might be there is some kind of number that is not rational and not irrational. Logicians refer to this as the problem of the excluded middle. Of course, if by irrational we mean not rational, then the problem of the excluded middle goes away. What about numbers written in base nine? Our proof says nothing about numbers written in base nine. In base seven, the square root of 12 is 3, $\sqrt{12_7} = 3$; of course $12_7 = 9_{10}$. If we accept that choice of number base is arbitrary, then we accept that any number which is not rational when written in base ten is not rational when written in base nine.

We see that even a proof as simple as (1.1) has to be thought about carefully. How do we know that, a hundred years from now, a five-year-old pupil at some infant school will not find a fault in the proof above, (1.1)? We do not know, but we are 100% confident that there is no fault in the above proof. Such is the mathematicians claim to hold absolute truth. Surely, it is impossible for mathematicians be mistaken?

About this book:

It is the opinion of your author that there are errors within modern mathematics and modern physics. It is the opinion of your author that great individuals in both mathematics and physics have made errors which continue to be promulgated today. This book will look at twelve such alleged errors; we will also have a brief look at some petty careless errors. These twelve alleged errors are probably not the only errors in modern mathematics, but they are alleged errors that have substantial implications. The alleged errors we look at in this book are:

1) The algebraic extension error
2) The hyperbolic complex numbers error
3) The all rotations are 2-dimensional error
4) Oersted and the breakdown of parity error
5) The rotation about an axis error
6) The error of differentiation in $\mathbb{C}$
7) The empty space error
8) The error of Hilbert spaces
9) The error of Pauli's spinors
10) The errors in the Dirac equation
11) The scattering amplitude error
12) The Lie group error

Introduction

We will look at the mathematics of the alleged errors, but we will also examine the historical and social milieu surrounding the making of these alleged errors. It will transpire that the errors are not due to carelessness or stupidity. There are various reasons these errors arose, and there have been various consequences to these errors, not all of which were bad. It seems that, in mathematics and physics, as in many aspects of life, it is often better to err than it is to do nothing.

As the reader will see, we are challenging the standard mantra produced by several towering intellects of mathematics and physics, but before we get into the details, we will have a general look at the nature of mathematical proof and of mathematics.

Chapter 2

The Nature of Mathematical Proof

The social side of mathematical proof:
Mathematical proof has a social side to its nature. A mathematician will present what she believes to be a proof of a mathematical theorem to the world. Any person in the world is free to view the alleged proof and to search for errors within that presented proof. If no-one can find any errors in the proof, then the proof is taken to be correct.

However, it needs only one person to find an error in the alleged proof for that alleged proof to be dismissed as being incorrect. In reality, the finder of the error has to shout quite loudly or has to be of considerable status within the mathematical profession, but sooner or later, the finder of an error will be heard. We give some examples of this social nature of mathematical proof.

Theorema egregium:
Our first example concerns the mathematician Gauss. In 1828, Gauss, often called the 'king of mathematicians' and generally ranked as being the greatest mathematician that ever lived[1], presented to the world a proof of what he called *theorema egregium*[2]. Translated, *theorema egregium* means 'king of theorems'. Gauss gave this name to his theorem because he, 'the king of mathematicians', thought the theorem so important. The theorem says that an empty space can be (intrinsically) curved without having to be sitting in a higher dimensional space.

---

[1] Such rankings are entirely subjective. By the way, Bertrand Russell is ranked second and either Euler or Newton are ranked third depending upon who you ask.
[2] Presented in 'Disquisitions Generales circa Superficies Curvas' by Carolo Friderico Gauss.

The surface of a sphere is curved within the 3-dimensional spatial part of our 4-dimensional space-time, but according to *theorema egregium*, the sphere does not need to be sitting in a higher dimensional space to be curved. This theorem underpins the theory of general relativity which treats our 4-dimensional space as being curved but not as being within a higher dimensional space.

When Gauss presented his proof of *theorema egregium*, it was perused by thousands of contemporary mathematicians and not one of these thousands of contemporary mathematicians could find any error in Gauss's proof. Thus, *theorema egregium* was pronounced proven.

We now know of four distinct errors in Gauss's proof. The errors are of a topological nature and would not have been realised by mathematicians alive at the time Gauss published. The errors have been corrected, and *theorema egregium* still stands. The lesson of this is that, if even the 'king of mathematicians' can make an error, especially in the 'king of theorems', then anyone can make an error; all of Gauss's contemporaries failed to see that error.

In the case of *theorema egregium*, the four discovered errors did not mean that *theorema egregium* was incorrect; they meant only that *theorema egregium* had not been proven. Often, much mathematics and some physics is built upon a mathematical theorem. If the theorem is incorrect, this incorrectness would most likely manifest itself as inconsistencies in the mathematics and physics built upon the theorem.

We can almost say that the observational proof of the theory of general relativity is also observational proof of *theorema egregium*. The *theorema egregium* must be true if general relativity is true. Of course, *theorema egregium* might still be true even if general relativity is not true.

Fermat's last theorem:
Our second example concerns the contemporary mathematician Andrew Wiles (1953-). On 23rd June 1993, Andrew Wiles presented his

proof of Fermat's last theorem[3] to the world[4]. In September of that year, Wiles' proof was found to contain an error. The error was very subtle; it had to be subtle for someone like Wiles to miss it. The error was so subtle that thousands of mathematicians who examined Wiles' proof also missed the error. The error was found by a single mathematician who specialised in one of the very many areas of mathematics which Wiles had needed to master to produce his proof.

It took Wiles a year to correct the error. He announced the correction on 19th September 1994. The correction required a deep insight which Wiles described as "… indescribably beautiful …" which came to him as "… The most important moment of his working life. …". The fully correct proof was published in May 1995; it is 150 pages long.

Wiles' proof is 150 pages of highly specialised mathematics involving many areas of mathematics. Only very highly trained and highly specialised mathematicians can understand all the proof. How can we be sure there is not another error in the proof that has been missed by all mathematicians? If Wiles can miss an error, anybody can. We cannot be sure.

Complicated proofs:
In the introduction, we gave a very simple proof that numbers, written in base ten, which have a final digit as $\{2, 3, 7, 8\}$ cannot have rational square roots. We feel very confident that there is no error in this proof because we can see the whole of the proof very clearly – it is simple.

Above, we mentioned the proof of Fermat's last theorem by Andrew Wiles. This is far from a simple proof, but it is understood completely by at least one person, Wiles. We will give another example of a complicated proof; this proof is completely understood by no single individual. This other example is 'The Classification of the Simple

---

[3] Wiles actually presented his proof of the modularity theorem for semistable elliptic curves. This theorem, together with Ribet's theorem, effectively proved Fermat's last theorem.
[4] Wiles gave the proof in a lecture 'Elliptic Curves and Galois Representations' at Cambridge University.

Finite Groups Theorem'; this is generally known as the 'Classification Theorem'.

The Classification theorem:
There are 'lots of infinities worth' of finite groups. Within this 'lots of infinities worth' of finite groups, there are an infinite number of special types of finite groups called 'the simple finite groups'. A simple finite group in group theory is similar to a prime number in number theory.

Beginning circa 1870[5], mathematicians like Camille Jordan (1838-1922) began to classify all the finite simple groups. The classification was completed 134 years later in 2004 by the mathematician Michael Aschbacher[6] (1944-). The simple finite groups are classified into four types[7] and 26 individual sporadic simple groups. The four types are the cyclic groups of prime order, the alternating groups of order 60 and above, and two types of classical Lie groups.

Aside: It is remarkable that, in something as orderly as finite group theory and mathematics in general, there should be 26, and exactly 26, individual odd balls.

The proof of the Classification Theorem is spread over 134 years and over tens of thousands of pages in several hundred journal articles produced by approximately one hundred different authors. Some of the articles which comprise this proof are themselves over a thousand pages long. It is said that no living mathematician understands the whole proof, and dead mathematicians certainly do not understand it.

Doubtless mathematicians have tried their best to ensure there are no errors in the proof of the Classification theorem, but the chance of zero errors does seem remote. In fact, errors are found in the proof of the Classification theorem and corrected almost daily.

---

[5] Jorden listed some simple finite groups in 1870. By 1893, Cole had classified the finite simple groups up to order 660.
[6] Michael Aschbacher : The Status of the Classification of the Finite Simple Groups : Notices of the American Mathematical Society 51(7) pg 736-740.
[7] Some people count differently and get seven distinct types.

In practice, mathematicians will use the Classification theorem until it 'causes' a noticed substantial error in some other area of mathematics. At that time, they will seek out the cause of the error and perhaps trace it back to an error in the proof of the Classification theorem. What more can you do? You can comb through the entire proof looking for errors and trying to understand the proof more clearly in the hope of being able to reduce the size of the proof to something more manageable. This is exactly what is being done by many dedicated mathematicians at this time.

Underpinning this effort to bring the Classification theorem under control, there is the assumption that a good mathematical proof should be 'quick and simple'. The proof of the Classification theorem might be so long because we have missed a profound insight; if we can find that profound insight, we opine that the proof of the Classification theorem will become 'quick and simple'.

Summary of social side of mathematical proof:
Looking at the above three examples, we see that mathematics is not the ivory tower of certainty that it is often presented as being. Mathematicians might pull out their hair and curse us, but, quite simply, the mathematician's 'absolute truth' is not absolute.

We are still left with the proof of the irrationality of the square root of 2 given above, (1.1). This proof surely cannot be in error. How about:

$$\begin{bmatrix} 0 & 1 \\ 2 & 0 \end{bmatrix} \begin{bmatrix} 0 & 1 \\ 2 & 0 \end{bmatrix} = \begin{bmatrix} 2 & 0 \\ 0 & 2 \end{bmatrix} \qquad (2.1)$$

Of course, the given square root is not an element of a division algebra (not a number).

Perhaps an alternative proof of the irrationality of the square root of two would be the simple statement, "John Peters says so".[8]

Undeclared assumptions:

The proof of the irrationality of the square root of 2 given above, (1.1), assumed that the square root of 2 had to be a real number. This seems to be a reasonable assumption, but we did not declare it, and we did not realise we were making this assumption.

Prior to the advent of quantum physics, physicists had unquestioningly assumed the universe was deterministic; it seemed impossible for the universe to be otherwise. We now know that the universe is not deterministic. From 1932 until immediately prior to the discovery of parity violation by the weak force, physicists had unquestioningly assumed the universe did not prefer its' left-hand over its' right-hand. We now know that the universe does prefer its' left-hand over its' right hand.

Most mathematicians have experienced producing what seemed to be a perfect mathematical proof only to discover that they had made an unrealised assumption and the proof will not stand. Often, the unrealised assumption is later pointed out by some other mathematician. It is very difficult to consciously know exactly what one is assuming in any human endeavour. Mathematics is the same.

Undeclared assumptions of Riemann geometry:

Riemann geometry assumes that n-dimensional space is comprised of n copies of the 1-dimensional real numbers. Such spaces are referred to as $\mathbb{R}^n$. Riemann geometry further assumes that all the angles within a space are 2-dimensional angles that 'pop into existence' in just the

---

[8] John Peters was a long suffering mathematics tutor of your author. Your author, stuck for a proof in an exam, once gave the John Peters proof in lieu. The examiner, John Peters, gave your author only one mark for this excellent proof.

needed number simply because there are some 1-dimensional spaces around.

We now know that there are 3-dimensional angles and 4-dimensional angles and, indeed, angles of any dimension. We now also know that there is a 4-dimensional space which has only two angles rather than six.[9] With this knowledge, we can question the 'angles pop into existence' assumption. Prior to having this knowledge, it occurred to no one that angles could be other than 2-dimensional or that a space might not have angles between all its axes.

One of the dangers of any area of mathematics is that it might be underpinned by undeclared assumptions which no-one has yet noticed even though, once seen, these assumptions are obvious. It is quite obvious that we need angles to form a geometric space just as much as we need axes, but modern mathematics forms a space with only axes; the notation, $\mathbb{R}^n$ is a clear statement of this mantra.

Another undeclared assumption:
Consider a plate spinning horizontally on top of a vertical rod in a circus act. What kind of rotation is the plate exhibiting? Think for a while. It is exhibiting 2-dimensional Euclidean rotation. Great, but you've missed the other two types of rotation being exhibited. As a point on the edge of the plate rotates, it accelerates and decelerates in two orthogonal horizontal directions. Acceleration is change of velocity; change of velocity is rotation in 2-dimensional space-time. The spinning plate is rotating in three 2-dimensional planes; these are one Euclidean plane and two hyperbolic space-time planes.

Can we rotate in the complex plane, $\mathbb{C}$? There is no space-time in the complex plane. Wiley these undeclared assumptions!

---

[9] This is the quaternion emergent expectation space of the electron. The lack of sufficient angles to support unfettered rotation is why electron spin is directionally quantitised.

Computer proofs:

Some mathematical theorems are proven by letting a computer do an immense number of calculations testing every possible case in which a theorem might be wrong. Attempts to do this by human hand are error prone and would often take an entire lifetime. The 4-colour map theorem was proven by computer.

The 4-colour map theorem states that, if you have a flat Euclidean plane divided into separate regions of any shape just like a map of countries, then four distinct colours are a sufficient number of colours to enable you to colour each of the regions in such a way that no two regions sharing a border are the same colour. This theorem was notoriously difficult to prove.

The 5-colour map theorem is easily proved; it was proved in 1890 by Percy John Heawood[10] (1861-1955) of Durham University. At the same time as proving the 5-colour map theorem, Heawood found an error in a previous proof of the 4-colour map theorem which had been presented in 1879 by Alfred Kempe (1849-1922) and had been accepted as a correct proof by the mathematics community for the previous eleven years.

After a great deal of effort, Kenneth Appel (1932-2013) and Wolfgang Haken (1928-) proved the 4-colour map theorem in 1976 by using a computer to examine 1,936 individual maps[11].

The proof by computer was rejected by most mathematicians in the 1970's – how do you know that you have not made a programming error? Today, most mathematicians accept proof by computer, but some still refuse. There is a theorem of logic which states that, no matter how small or how simple your computer program might be, you cannot be sure there are no 'bugs' in that program. Thus, in accepting computer proofs, mathematicians are surrendering their claim to be the holders of absolute truth.

---

[10] Heawood P. J. (1890) Map Colour Theorem : Quarterly Journal of Mathematics Oxford 24 pg 332-338

[11] Appel, Kenneth. Haken, Wolfgang Every Planar Map is 4-colourable : Illinois Journal of Mathematics 21(3) pgs 429-490 & 491 - 567

Looking back to the proof given above that the square root of two is irrational, (1.1), we see that this is really a computer proof. In this case, the human brain is the computer. Perhaps, as a computer running a program, the human brain has 'bugs' in it; if so, then the purported proof, (1.1), might not be a proof even though it appears to be a proof to us.

Perhaps our mind is wired, we might say programed, in such a way that we perceive proofs where there are none or that we derive proofs relating to mathematical objects which do not really exist but are merely the products of our minds being wired in a particular way. The philosopher Immanuel Kant (1724-1804) was of the opinion that our concept of space and time are no more than figments generated by the way our minds are wired. If Kant is correct, Riemann geometry, and the *theorema egregium*, is the mathematics of something that does not really exist.

The Goldbach conjecture:
The Goldbach conjecture, named after Christian Goldbach (1690-1764) who presented the conjecture to Leonhard Euler (1707-1783) on 7th June 1742, states that every even integer greater than four can be expressed as the sum of two odd prime numbers; for example:

$$18 = 7 + 13$$
$$20 = 3 + 17$$
(2.2)

This conjecture has been proven to be true by computer up to the number 10^{18}, but it is still a conjecture. There might be an even number greater than 10^{18} which cannot be expressed as the sum of two odd primes. It is a fact that prime numbers occur less often as we count higher along the real number line; perhaps this scarcity of high prime numbers will eventually make the Goldbach conjecture not true.

Having tested all even numbers up to 10^{18}, the reader might take the view that the chance of the Goldbach conjecture being wrong is less than the chance of thousands of mathematicians having missed an error in Wiles' proof of Fermat's last theorem. If we accept Wiles' proof of

Fermat's last theorem, why do we not accept this statistically more certain computer proof of the Goldbach's conjecture?

Proof is something more than proof:
Within mathematics there are 'good proofs' and there are 'poor proofs'. The reader might wonder how one proof can be better than another proof if they both prove the same thing.

The proof we gave above that the square root of two is not rational, (1.1) is a 'good proof'. The Classification Theorem of Simple Finite Groups is a 'poor proof'. The difference is subjective.

Obscurity is a bad thing to have within a proof; it usually hides unrealised assumptions. Simplicity is a good thing in a proof. The Nobel Laureate physicist Richard Feynman once said, "If you understand something properly, you should be able to explain it to a 12-year old". A good proof is one that can be explained to a 12-year old.

Some proofs contain a great insight within them. Wiles' first proof of Fermat's last theorem (the incorrect one) lacked the "... indescribably beautiful ... most important moment of his working life. ..." which is in Wiles' second (the correct one) proof of Fermat's last theorem. A great insight is associated with a 'good proof' partly because it simplifies the proof but also because a great insight is of value in itself. Of course, Fermat claimed to have proved Fermat's last theorem. Perhaps Fermat had a great insight that enabled him to prove his last theorem with no more than a few simple lines. If so, then Wiles' proof is a 'poor proof' whereas Fermat's proof would be a 'very good proof'.

In short, mathematicians want more from a proof than just a proof. They want a deeper understanding of why the proven theorem is true. If a proof provides that deep understanding, it is a good proof.

There is an Open University problem within the course on programming with the Maple mathematics computer program which requires the student to write a program which successively adds all the digits of a number until we are left with only one digit; such a sum is called the digital root of the number; for example:

$$DR[1234567897] = DR[1+2+3+4+5+6+7+8+9+7] \\ = DR[52] = DR[5+2] = 7 \qquad (2.3)$$

The number 7 is the digital root of 1234567897. The idea behind the problem is that the student must use while loops. An easier way to do the calculation is by simply taking the remainder after dividing the number by 9 (this is the number base minus one):

$$\frac{1234567897}{9} = 137174210 \quad remainder \quad 7 \qquad (2.4)$$

The complicated program with loops is far more likely to contain 'bugs' than the divide by nine program, but, more than this, the divide by nine program 'understands' the problem at a much deeper level than the loop program.

Summary:

Mathematical proof is not absolutely guaranteed to be correct. Mathematicians make errors which are not noticed by their contemporaries. Mathematicians unwittingly make undeclared assumptions when constructing an area of mathematics. Computer programs can never be declared to be 'bug free' with certainty.

Mathematics is not the ivory tower of absolute truth which it is often claimed to be. Mathematicians need to be humble like physicists.

Chapter 3

Two Ways to Construct Mathematics

Modern mathematics is a huge amount of human knowledge. No single mathematician can know all modern mathematics.

The whole of this immense construction which is modern mathematics is self-consistent. There are no contradictions in mathematics. Everything holds together solidly.

The same is true of a good fairy story. A good fairy story is self-consistent, without contradictions, and holds together solidly.

The axiomatic system:
Modern mathematics is built upon thirty four axioms which are declared to be absolutely true. There are also four logical axioms which dictate how one can build mathematics from the axioms. All these axioms are entirely self-consistent, and so, provided we build mathematics on no more than these axioms in a way entirely consistent with the logical axioms, we will have no inconsistencies or contradictions in what we build. This is precisely what has been done by mathematicians over the last two millennia.

Initially, circa 300 BC, Euclid started this building process with just five declared axioms and some unrealised and undeclared axioms[12]. Additional axioms have been carefully added over the centuries. Great care is taken to ensure that a new axiom is consistent with the established axioms. Mathematical proofs have been added to the mathematics very carefully over the centuries.

---

[12] Not much is known about Euclid's life, but most historians opine that he was born around 330 BC to wealthy parents in Alexandria; other historians opine that he was born in Tyre but spent his life in Damascus.

The care that has been taken from Euclid onward is such that today's mathematicians can confidently claim to have built a system of knowledge that is without internal contradictions and is soundly self-consistent.

The mathematics that has been built has found widespread application within engineering, physics and the other sciences. The axiomatic system has been very useful to humankind.

Starting with an axiomatic system is exactly the correct way to formulate a good fairy story that will hold true and be believable to its readers.

Deduction from the number one:
Perhaps the axiomatic system is not the correct way to produce mathematics. Perhaps we should start with no-more than the number one and from that deduce mathematics using logical deduction. How far would we get?

We would get remarkably far. We would get general relativity and classical electromagnetism. We would get electrons, neutrinos and the weak force. We would get our unique expanding universe with its inflationary beginning. We would get special relativity. We would get gauge theory. We would get two 2nd rank tensors. If fact, it seems that we might eventually get the whole of physics.

We would get complex numbers and quaternions. We would get real numbers and finite groups. We would get Riemann geometry but in only one space, our 4-dimensional space-time. We would get calculus[13]. We would even get Clifford algebras. We know we would get all these things because we already have them.

We would not get 5-dimensional Riemann space. There are other things which we opine we would not get, but it is hard to be sure. We opine that we would get enough mathematics to give us the whole of physics

---

[13] Calculus within the complex numbers would be different to the Cauchy-Riemann calculus conventionally used within the complex numbers.

and some more but that we would get only enough mathematics to give us a small part of modern conventional mathematics.

Your author is not the first person to advocate building mathematics from only the number one. In the first years of the 20th century, Bertrand Russell showed that if the number one exists, then so must all the other real numbers and that one plus one equals two. Russell showed not only the existence of the real numbers but also that they were held together as an algebra by the algebraic operations of addition and multiplication. This took Russell fifteen years. It is by far the hardest part of building mathematics from no more than the number one. Riding on the back of Russell, the rest is comparatively easy.

Two ways of building mathematics:
We see that there are at least two radically different ways of building mathematics.

1) The axiomatic system
2) From no more than the number one

The axiomatic system will allow the construction of much mathematics which does not really exist. Fairy stories can be internally consistent, but that does not mean that fairies exist. An example of this is 5-dimensional Riemann space; this is easy to construct, but we are now sure that it does not exist.

If the existence of number one leads to some mathematical entity, a set of trigonometric functions for example, then we can be sure that entity really exists because the number one really exists and the given mathematical entity is consequent to the existence of the number one.

As the reader might have realised, your author prefers building mathematics from no more than the existence of the number one over building mathematics using the axiomatic system. It seems to your author that much of modern mathematics is invented rather than discovered. This invented mathematics is consistent with all other mathematics, but it is still invented. Grimm's fairy stories are consistent with each other, but fairies do not exist.

We might assert that using the axiomatic system is a huge error. Perhaps future generations will take that view, but, let us recall the immense progress that humankind has made using the axiomatic system. If it was an error, it was a grievously good one. However, perhaps it is time for change.

Introduction to the Ensuing Chapters

With the next chapter, we begin our consideration of some alleged errors in modern mathematics. These alleged errors are of various types and arise for various reasons. We do not seek to denigrate the mathematicians who made these errors or who continue to promulgate these errors. As will become apparent, often, the making of the error was the only way to progress at the time the error was made.

We bear in mind that we have the advantage of hindsight. Doubtless, if we had been in the place of the mathematicians who originally made these errors, we too would have made these errors.

Scientific advance is often portrayed as an efficiently walked straight road leading from ignorance to enlightenment. Along this road we picture the occasional bright lights of some great insight made by some great intellect. The reality is that scientific advance is more like a blind man stumbling and staggering along an unmapped and unlit country lane on a moonless night. He spends more time tangled in brambles or sinking in sodden marshland than he spends moving forward along the road. The bright lights we see with hindsight as shining beacons when we look back along the road we have walked are more often the product of luck than of great intellect, or, at least, luck played a large part in lighting them, luck and the obstinacy of the person who first lit them.

Homo sapiens are not as sapient as we often see ourselves to be.

Chapter 4

The Algebraic Extension Error

Some 500 years ago, a major interest of mathematicians was solving polynomial equations. The way to solve polynomial equations is to factorise them into linear factors. For example:

$$x^2 + x - 12 = 0$$
$$(x+4)(x-3) = 0 \qquad (4.1)$$

This polynomial equation has the solutions:

$$x + 4 = 0$$
$$x - 3 = 0 \qquad (4.2)$$

Some 500 years ago, mathematicians knew of only the real numbers, and, not knowing of complex numbers, they sought solutions to polynomial equations which were real numbers. In other words, mathematicians sought to factorise the polynomials into linear factors using only real numbers. Of course, there was interest in far more complicated polynomials than the above, (4.1), and mathematicians had developed methods for solving quite complicated polynomials. However, there were some polynomials, even quite simple ones, which could not be factorised into linear factors using only real numbers, for example:

$$x^2 + 1 = 0 \qquad (4.3)$$

This, (4.3), is called a monic minimum polynomial. It is monic because the coefficient of the highest power of the variable, x^2 in this case, is one. It is minimal because it will not factorise within the number system (division algebra) of the coefficients, the real numbers in this case.

To factorise this polynomial equation, (4.3), requires a number that is a square root of minus one, but there is no real number which is a square root of minus one. If we are to have solutions to this polynomial equation, (4.3), it is necessary to extend the real numbers to include a square root of minus one.

$$(x-\sqrt{-1})(x+\sqrt{-1})=0 \qquad (4.4)$$

It was Gerolamo Cardano (1501-1576) who first introduced a square root of minus one to the world in his book *Ars Magna* published in 1545.[14] Effectively, by adding the square root of minus one into the real numbers, Cardano extended the 1-dimensional real numbers into what we now call the 2-dimensional complex numbers, $\mathbb{C}$. Of course, it took mathematicians many years to understand that this is what Cardano had done.

The English mathematician John Wallis (1616-1703) was the first to associate Cardano's square root of minus one with the 2-dimensional space we now call the complex plane. He did this in 1673. Although Wallis's association was repeated by many mathematicians, it was Jean-Robert Argand (1768-1822) who eventually got the credit for Wallis's association in 1806 when the complex plane was named the Argand diagram; that is 261 years after Cardano's publication of *Ars Magna*.

None-the-less, the stage was set by Cardano and Wallis. We have 1-dimensional numbers, the real numbers. We have 2-dimensional numbers, the complex numbers. There might be 3-dimensional numbers. How do we find them? At this point, following Wallis, mathematicians took a wrong road; in their position, we would have doubtless taken the same wrong road.

The wrong road:
Mathematicians of the 17th and 18th centuries sought higher dimensional types of complex numbers by seeking polynomials that will not factorise into linear factors using only the 2-dimensional

---

[14] Other mathematicians involved include Niccolo Foantana Tartaglia and Lodovico Ferrari.

complex numbers. They sought what is now called an algebraic extension of the complex numbers, $\mathbb{C}$, based on a monic minimum polynomial.

We now know there are no polynomials with complex coefficients and complex variables which will not factorise within the 2-dimensional complex numbers; that is, we know there is no monic minimum polynomials that cannot be factorised into linear factors within the complex numbers. That all complex polynomials within the complex numbers can be factorised is called the algebraic closure of the complex numbers.

Consequences of the error:
Following the accepted realisation by Wallis and others that higher dimensional complex numbers are associated with monic minimum polynomials which cannot be factorised within a particular number system (division algebra), there was a very considerable effort by many mathematicians to seek 3-dimensional complex numbers by seeking polynomials that cannot be factorised within the 2-dimensional complex numbers. Since we live in what was then thought to be 3-dimensional space, such numbers must surely exist. When, after great efforts, the searching for 3-dimensional complex numbers proved fruitless, even greater effort went into attempting to prove that 3-dimensional complex numbers do not exist. This effort was spread over many years and many mathematicians until the algebraic closure of the complex numbers was eventually proven by Gauss and published in 1799[15].

The algebraic closure of the 2-dimensional complex numbers means that all polynomials with complex coefficients can be factorised into linear factors within the 2-dimensional complex numbers. With no more polynomials that will not factorise into linear factors, there is seemingly no way to extend the 2-dimensional complex numbers into 3-dimensions. Thus, the algebraic closure of the 2-dimensional

---

[15] This was Gauss's doctoral thesis; he was aged 19 at the time he wrote it.

complex numbers was taken at the time to prove that 3-dimensional complex numbers do not exist.

This association between higher dimensional complex numbers and polynomials which will not factorise into linear factors was an error which had considerable consequences within the history of human scientific development. Yet see how easy it is to go down this road. Do you see how this road was obviously the way forward? Furthermore, 300 years ago, there was no other road in sight.

Aside: The mathematician Joseph-Louis Lagrange (1736-1813) apparently produced a proof in 1830 (17 years after his death) that there are no 3-dimensional complex numbers. The proof is simply the fact that the product of two 3-dimensional quadratic forms is not a 3-dimensional quadratic form. In mathematical notation, this is:

$$(x^2 + y^2)(a^2 + b^2) = (ax - by)^2 + (bx + ay)^2$$
but (4.5)
$$(x^2 + y^2 + z^2)(a^2 + b^2 + c^2) \neq P^2 + Q^2 + R^2$$

wherein we have presented the 2-dimensional quadratic form which is the complex numbers, $\mathbb{C}$, for comparison. In technical parlance, a division algebra must have '*norm of product equals product of norms*'. Since the 3-dimensional quadratic form does not have this property, there can be no 3-dimensional complex numbers. Well! there are no 3-dimensional complex numbers which include the 2-dimensional complex numbers as a sub-algebra, and this is what Lagrange proved. We now know that there are 3-dimensional complex numbers, but they do not have the 2-dimensional complex numbers as a sub-algebra. It is remarkable that Lagrange was able to prove this 17 years after he died.

Of course, after the proof of algebraic closure by Gauss, everyone stopped looking for higher dimensional complex numbers. Even today, many mathematicians will tell you that it has been proven that there are no 3-dimensional complex numbers[16].

---

[16] News travels slowly.

More history:

Remarkably, in 1843, the Irish mathematician William Hamilton (1805-1865) discovered the 4-dimensional quaternions. Gauss's proof of the algebraic closure of the complex numbers is correct, and so the existence of the quaternions indicates a disassociation of higher dimensional complex numbers and monic minimum polynomials. Two months later, John Thomas Graves (1806-1870) discovered the 8-dimensional octonians.

In 1848, James Cockle (1819-1895) discovered the 2-dimensional hyperbolic complex numbers. This was another nail in the coffin of the algebraic extension viewpoint.

It is remarkable that even today most of the mathematical fraternity still associates the complex numbers with monic minimum polynomials and algebraic extension and thereby denies the existence of all the higher dimensional complex numbers except the quaternions which tend to be swept under the carpet on the grounds that, "They are not proper numbers because they are non-commutative". We still often read that the complex numbers, $\mathbb{C}$, are an algebraic extension of the real numbers, $\mathbb{R}$, based on a monic minimum polynomial. This is the standard mantra within abstract algebra.

The error explained:

Why is it in error to seek higher dimensional types of complex numbers by seeking polynomials which will not factorise into linear factors using only the 2-dimensional complex numbers?

We now know that the 2-dimensional complex numbers are only coincidently associated with monic minimum polynomials like (4.3). The complex numbers actually derive from the order two finite group C_2. We came to understand this in only 2007[17]. The real numbers derive from the order one finite group C_1.

---

[17] See: Dennis Morris : Complex Numbers The Higher Dimensional Forms 2nd Edition.

The Algebraic Extension Error

Prior to 2007, all mathematicians considered the 2-dimensional complex numbers, $\mathbb{C}$, to be an algebraic extension of the real numbers. Even today, nine years later, the news that they derive from the order two finite group C_2 has not yet spread to the majority of the world's mathematicians. It will be fifty years before the mathematics fraternity at large is aware of the fact that the 2-dimensional complex numbers, $\mathbb{C}$, derive from the C_2 finite group.

With hindsight, it is obvious that the complex numbers derive from a finite group. Any division algebra satisfies thirteen division algebra axioms. Of these thirteen axioms, four are the axioms that define a finite group. Every division algebra has a finite group at its heart. We will see this shortly.

It takes only a little thought to realise that the 1-dimensional real numbers derive from the C_1 finite group. So where might we find the 3-dimensional complex numbers? The reader knows the answer; we expect to find the 3-dimensional complex numbers in the order three finite group C_3. Similarly, there are higher dimensional complex numbers within every finite group. Quite remarkably, the multiplicatively non-commutative quaternions derive from the commutative $C_2 \times C_2$ finite group. Cockle's hyperbolic complex numbers also derive from the C_2 finite group[18].

You see that the finite group road we are about to walk is a road very different from the algebraic extension road walked by our forebears. We have pulled ourselves from the sodden marshland back on to the correct road. No doubt we will walk it as far as we can, but, sooner or later, we will wander from it into some ditch or bramble patch.

The finite group road:
How do we know that the finite group road is the correct road to follow? Well, other than the obvious fact that every division algebra has a finite

---

[18] The octonians are not a division algebra and do not derive from any finite group.

group at its heart, the answer is by results. Following the finite group road is not only simpler but has produced a huge spurt forward in mathematics and is currently expected to provide the basis of a rewrite of theoretical physics that will be the long sought for grand unification. Mathematics has been revolutionised by following the finite group road, and physics is looking very much like it is about to be revolutionised[19].

Interim summary:
In 1545, and for centuries after, there were very understandable reasons for walking the algebraic extension road. No-one knew of group theory, and so there was no group theory road to walk. Clearly, the 2-dimensional complex numbers had emerged from a polynomial that could not be factorised. It all made sense.

However, there was a much more subtle and unrealised assumption made by mathematicians looking for 3-dimensional complex numbers. They assumed, without stating it, and probably without realising they were making this assumption, that 2-dimensional space is a sub-space of 3-dimensional space. That is true of our 4-dimensional space-time; it is not true of the spinor spaces (division algebra spaces). From this assumption, it follows that the 2-dimensional complex numbers must be a sub-algebra of the 3-dimensional complex numbers. This constrained the search for 3-dimensional complex numbers by the condition that they must be an extension of the 2-dimensional complex numbers. We now know that the 2-dimensional complex numbers are not a sub-algebra of the 3-dimensional complex numbers. In 1545, there was no reason to question that 2-dimensional space is a sub-space of 3-dimensional space, and so no reason to think that 2-dimensional complex numbers would not be a sub-algebra of 3-dimensional complex numbers.

This error was not an error made by mathematicians being foolish. This error was forced upon mathematicians by history. How much further

---

[19] Of course, it will take fifty years for the revolution to spread throughout the mathematics fraternity.

ahead might we have been today if mathematicians had realised that the complex numbers derive from the finite groups in 1546 rather than in 2007? Not very much further, finite groups were not discovered until the 19th century. Without knowing of finite groups, even the greatest mathematical genius could not have properly understood the genesis of the complex numbers in 1546.

Notation:

Large steps forward in mathematics are often associated with a change of notation. It might seem to the casual reader that notation is just cosmetic. An equation is the same equation written in Roman numerals as it is written in Arabic numerals. Well, it is, but you try doing number theory using Roman numerals. It is possible to do all mathematics using Roman numerals, but it would be very obscure and difficult mathematics to disentangle. There was a great flowering of number theory following the adoption of Arabic numerals to replace Roman numerals. Getting the right notation is often a major step in making progress in mathematics.

The normal notation for the real numbers and the 2-dimensional complex numbers is:

$$\begin{aligned} \mathbb{R}: &\quad a \\ \mathbb{C}: &\quad a + \hat{i}b \end{aligned} \qquad (4.6)$$

Let us write these types of numbers using matrix notation:

$$\begin{aligned} \mathbb{R}: &\quad [a] \\ \mathbb{C}: &\quad \begin{bmatrix} a & b \\ -b & a \end{bmatrix} \end{aligned} \qquad (4.7)$$

What might the 3-dimensional complex numbers look like in matrix notation? Even more suggestive is the permutation matrix notation for the finite groups:

$$C_1: \quad [1]$$

$$C_2: \quad \left\{\begin{bmatrix} 1 & 0 \\ 0 & 1 \end{bmatrix}, \begin{bmatrix} 0 & 1 \\ 1 & 0 \end{bmatrix}\right\} \quad (4.8)$$

It is not that we cannot use the established notation, (4.6), when dealing with complex numbers; we have used it for 400 years. It is that the matrix notation seems to 'know where to go next'. The finite group C_3 and one of the 3-dimensional complex number types is:

$$C_3: \quad \left\{\begin{bmatrix} 1 & 0 & 0 \\ 0 & 1 & 0 \\ 0 & 0 & 1 \end{bmatrix}, \begin{bmatrix} 0 & 1 & 0 \\ 0 & 0 & 1 \\ 1 & 0 & 0 \end{bmatrix}, \begin{bmatrix} 0 & 0 & 1 \\ 1 & 0 & 0 \\ 0 & 1 & 0 \end{bmatrix}\right\}$$

$$C_3 H^2: \quad \exp\left(\begin{bmatrix} a & b & c \\ c & a & b \\ b & c & a \end{bmatrix}\right) \quad (4.9)$$

Very capable mathematicians spent 300 years looking for the 3-dimensional complex numbers. We have just written one of them down in two lines. That is the power of good notation. Good notation goes hand-in-hand with understanding what we are doing. Often, good notation will convert an obscure mathematical 'poor proof' into a lucid 'good proof'.

There is even more to notation. A quaternion has traditionally been written as:

$$a + \hat{i}b + jc + kd \quad : \quad \hat{i}^2 = j^2 = k^2 = -1 \quad (4.10)$$

Setting any two of the imaginary variables to zero gives an expression like:

$$a + \hat{i}b \quad (4.11)$$

Within the traditional notation, this seems to be a complex number the same as (4.6), but it is actually a double cover of a complex number;

quaternion rotation is both clockwise and anti-clockwise at the same time. The matrix notation shows this clearly:

$$\begin{bmatrix} a & b & 0 & 0 \\ -b & a & 0 & 0 \\ 0 & 0 & a & -b \\ 0 & 0 & b & a \end{bmatrix} \neq \begin{bmatrix} a & b \\ -b & a \end{bmatrix} \qquad (4.12)$$

The double cover manifests itself physically as electrons being spin one-half particles. Without spin one-half particles, there is no universe, and so, by using traditional notation, we have lost the universe. Poor notation can hide things of importance.

The mathematics:
We have seen above, (4.3) & (4.4), how Cardano was led to the 2-dimensional complex numbers through a polynomial which could not be factorised using only real numbers. We will now derive the 2-dimensional complex numbers from the order two finite group C_2. We have presented C_2 above, (4.8), as a set of two permutation matrices. We take these two matrices, multiply each by a real variable and add them to get:

$$C_2 \rightarrow \begin{bmatrix} a & b \\ b & a \end{bmatrix} \qquad (4.13)$$

This is just the standard form Cayley table[20] of the finite group C_2.

Adding and multiplying two such matrices, (4.13), together, we discover that matrices of this form have all the properties we associate with numbers - they satisfy the axioms of a division algebra - except sometimes, when $a = b$, the matrix is singular and so it has no

---

[20] The standard form Cayley table is the Cayley table with all the identities on the leading diagonal.

multiplicative inverse. We can solve this by taking the exponential of the matrix (4.13). This gives:

$$\exp\left(\begin{bmatrix} a & b \\ b & a \end{bmatrix}\right) = \begin{bmatrix} h & 0 \\ 0 & h \end{bmatrix}\begin{bmatrix} \cosh b & \sinh b \\ \sinh b & \cosh b \end{bmatrix} \quad (4.14)$$

This matrix, (4.14), satisfies all the axioms of a division algebra except additive inverse on the real axis. If we wish, this is easily cured by putting a ± before the matrix but now-a-days we do not bother. Nothing is to be gained by the arbitrary insertion of the ± sign.

What we have here, (4.14), is the Lorentz transformation of special relativity:

$$\begin{bmatrix} \cosh b & \sinh b \\ \sinh b & \cosh b \end{bmatrix} = \begin{bmatrix} \gamma & v\gamma \\ v\gamma & \gamma \end{bmatrix}$$

$$\gamma = \frac{1}{\sqrt{1-\dfrac{v^2}{c^2}}} \quad (4.15)$$

These are the hyperbolic complex numbers, $\mathbb{S}$, first discovered by Cockle in 1848 – 57 years before Einstein presented special relativity.

In fact, playing with the axioms shows that the Cartesian matrix form of the hyperbolic complex numbers is[21]:

$$\mathbb{S}_\lambda = \exp\left(\begin{bmatrix} a & b \\ \lambda b & a \end{bmatrix}\right) \quad : \lambda > 0 \quad (4.16)$$

We do not require that $\lambda = 1$ to have the hyperbolic complex numbers; we require only that $\lambda > 0$. When $\lambda < 0$, we have the Euclidean complex numbers. We set $\lambda = -1$ for ease:

$$\mathbb{C} = \begin{bmatrix} a & b \\ -b & a \end{bmatrix} \quad (4.17)$$

---

[21] The proof of this is by demanding that the algebraic matrix form satisfies the thirteen division algebra axioms.

The Algebraic Extension Error

This is the complex numbers. The exponential, polar form, of this is:

$$\begin{bmatrix} r & 0 \\ 0 & r \end{bmatrix} \begin{bmatrix} \cos b & \sin b \\ -\sin b & \cos b \end{bmatrix} \quad (4.18)$$

And so you see how easy it is to derive the complex numbers and the other 2-dimensional division algebra from the finite group C_2.

There is more than just another type of numbers that has fallen from the finite group C_2. We have the two sets of 2-dimensional trigonometric functions within the two types of 2-dimensional rotation matrices. The whole idea of rotation, and with it angle, has fallen from the group C_2. Of course, two types of 2-dimensional angle have fallen out of the group C_2; these are the familiar Euclidean angles and the less familiar 2-dimensional space-time angles.

In spite of our education being such that we meet the (Euclidean) trigonometric functions years before we meet the complex numbers, these (Euclidean) trigonometric functions do not exist outside of the complex numbers, $\mathbb{C}$; the cosine and the sine are respectively the real and imaginary parts of a complex number. Historically, mathematicians have known of trigonometric functions since the 6th century[22] – a thousand years before we discovered the complex numbers.

The revolution:
The revolution in mathematics brought about by the above (4.8) to (4.18) is two-fold. Firstly, because there are an infinite number of finite groups, there is an infinite number of types of complex number (division algebras or spinor algebras), and, because there are finite groups of every order, there are types of complex number of every dimension. Secondly, the notation allows us to see right into the mathematics; the consequences of this are far-reaching and will lead to a complete revolution in our understanding of empty space.

---

[22] The sine and cosine functions were discovered by the Indian mathematician Aryabhata (476-550).

So, you want a 3-dimensional form of complex numbers, do not mess around looking for monic minimum polynomials that will not factorise; just write the order three group as a set of permutation matrices, multiply each permutation matrix by a real number and add these matrices to get the basic algebraic matrix form; then take the exponential of this matrix sum (algebraic matrix form) to be rid of singular matrices[23].

We note in passing, that since the order two finite group C_2 is not a sub-group of the order three finite group C_3, the 2-dimensional complex numbers are not a sub-algebra of the 3-dimensional complex numbers. You will not get the 3-dimensional complex numbers by extending the 2-dimensional complex numbers. For centuries, mathematicians were 'barking up the wrong tree'.

Addendum:

Consider the algebraic matrix form of the finite group C_3. Begin with the basic algebraic matrix form and multiply each element of the matrix by a parameter which is a real constant, $\alpha, \beta, \ldots$. The division algebra axioms demand that the parameter must be a real constant:

$$\begin{bmatrix} \alpha a & \beta b & \gamma c \\ \delta c & \varepsilon a & \chi b \\ \eta b & \kappa c & \mu a \end{bmatrix} \quad (4.19)$$

The constants on the top row, $\{\alpha, \beta, \gamma\}$, do no more than scale the variables. We can set them to one. A division algebra must have a multiplicative identity, and so the constants multiplying the a variable must be equal. This gives:

---

[23] A list of types of complex numbers up to order 15 is given in : Dennis Morris : The Uniqueness of our Space-time. For more details see: Dennis Morris : Complex Numbers The Higher Dimensional Forms.

$$\begin{bmatrix} a & b & c \\ \delta c & a & \chi b \\ \eta b & \kappa c & a \end{bmatrix} \quad (4.20)$$

We multiply two such matrices together and insist upon multiplicative closure of matrix form[24]. This leads to the elimination of some of the constants, and we get:

$$\exp\left(\begin{bmatrix} a & b & c \\ \delta c & a & \chi b \\ \delta b & \dfrac{\delta}{\chi} c & a \end{bmatrix} \right) \quad (4.21)$$

Choosing the four permutations $\delta = \pm 1$, $\chi = \pm 1$ will give the four types of 3-dimensional complex numbers; two of these are algebraically isomorphic.

In non-matrix notation, these 3-dimensional algebras can be written as:

$$\begin{array}{ll} \exp\left(a + b\sqrt[3]{+1} + c\sqrt[3]{+1}\right) & \exp\left(a + b\sqrt[3]{-1} + c\sqrt[3]{+1}\right) \\ \exp\left(a + b\sqrt[3]{+1} + c\sqrt[3]{-1}\right) & \exp\left(a + b\sqrt[3]{-1} + c\sqrt[3]{-1}\right) \end{array} \quad (4.22)$$

The C_4 algebras are all fourth roots of ± 1; the C_5 algebras are all fifth roots of ± 1, and so on for all the cyclic groups.

---

[24] See : Dennis Morris : The Physics of Empty Space or Dennis Morris : Complex Numbers The Higher Dimensional Forms.

Chapter 5

The Hyperbolic Complex Numbers Error

In the previous chapter, we mentioned the hyperbolic complex numbers first discovered by James Cockle in 1848. We reproduce these complex number here for the convenience of the reader:

$$\exp\left(\begin{bmatrix} a & b \\ b & a \end{bmatrix}\right) = \begin{bmatrix} h & 0 \\ 0 & h \end{bmatrix}\begin{bmatrix} \cosh b & \sinh b \\ \sinh b & \cosh b \end{bmatrix} \quad (5.1)$$

James Cockle was a mathematician of eminence. He was the president of the London Mathematical Society from 1886 to 1888.

The hyperbolic complex numbers, $\mathbb{S}$, are of algebraic status equal to the Euclidean complex numbers, $\mathbb{C}$. Both are algebraic fields. It is remarkable that although the Euclidean complex numbers are taught in every university on Earth, the hyperbolic complex numbers are, as far as your author can determine, taught at only Helsinki University and Beijing University.

After Cockle discovered the hyperbolic complex numbers, they were forgotten. How can something as important as a second type of 2-dimensional complex numbers be forgotten by mathematics? They have been rediscovered many times and often given different names. We have a list of the discoverers:

Tessarines	Cockle[25]	1848
Algebraic motors	Clifford[26]	1882

---

[25] Cockle James : On the symbols of algebra, and on the theory of tessarines Phil. Mag. (3), 34, 406-410

[26] Clifford, W.K. Mathematical Works (1882) pp 392 'Further notes on biquaternions'.

Hyperbolic complex numbers	Vignaux[27]	1935
	Sobczyk[28]	1995
	Guo Chun Wen	2002
Bireal numbers	Bencivenga[29]	1946
Approximate numbers	Warmus[30]	1956
Double numbers	Yaglom[31]	1968
	Hazewinkle[32]	1990
Anormal complex numbers	Benz[33]	1973
Perplex numbers	Fjelstad[34]	1986
Lorentz numbers	Harvey	1990
Dual numbers	Hucks	1993
Split-complex numbers	Rosenfield[35]	1997
Study numbers	Lounesto[36]	2001
Twocomplex numbers	Olariu[37]	2002

Perhaps we have missed someone.

---

[27] Sobre el numero complejo hiperbolico y su relacion con la geometria de Borel. Universidad Nacional de la plata Republica Argentina

[28] Sobczyk, G (1995) Hyperbolic number plane College Mathematics journal 26:268-80

[29] Sulla rappresentazione geometrica della algebra doppie dotate di modulo. Uldrico Bencivenga (1946)

[30] Calculus of Approximations. Bulletin de L'Academie Polonaise de Sciences (1956) Vol 4 No.5 pp 253-257

[31] Complex numbersin geometry. Academic Press N.Y. pp 18-20

[32] Double and Dual numbers – Encyclopedia of Mathematics Soviet/AMS/Kluwer, Dordrect

[33] Benz, W. Vorlesungen uber geometrie der algebren, Springer (1973)

[34] Extending Special Relativity with Perplex numbers. American Journal of Physics 54:416 (1986)

[35] Geometry of Lie groups, Kluwer academic publishers

[36] Clifford Algebras and Spinors ISBN: 0-521-00551-5

[37] Complex numbers in n-dimensions ISBN: 0-444-51123-7

Special relativity:

Even more remarkable is that the hyperbolic complex numbers division algebra is utterly central to the theory of special relativity. In some ways, this algebra is the special theory of relativity. One of the central concepts of modern physics is that any physical theory must be invariant under rotation in 2-dimensional space-time. This means that the theory complies with special relativity. The rotation matrix, (5.1), or rather the concept embodied in the rotation matrix, (5.1), is at the very centre of modern physics.

An error of forgetfulness:

It is an amazing aspect of the history of mathematics that the hyperbolic complex numbers have been discovered and forgotten so many times. Perhaps they were seen as uninteresting and unimportant prior to Einstein's publication of the special theory of relativity[38], this cannot be claimed since 1905.

It is hard to explain this error of mathematics. It is an error which continues to be promulgated today.

---

[38] Albert Einstein (1905) Zur Elektrodynamik bewegter Korper Annalen der Physik 17 891

Chapter 6

Careless Errors

This chapter is a lighter chapter than the previous two chapters. In this chapter, we look at a few commonly taught errors of physics which the teacher would not teach if they were a little more careful. Such errors circulate around the schools and universities until they become part of the 'common understanding' of physics. Consequently, this 'common understanding' is promulgated into the public domain by both academic media such as television documentaries or poorly written text-books and by school teachers who have accepted the 'common understanding' without realising the errors in that 'common understanding'.

Although experts in the particular areas would not make these errors, their expertise is often drowned by the cacophony of the 'common understanding'. The origin of these errors is not an academic failure of mathematicians or physicists. These errors are not unrecognised assumptions or errors due to researchers wandering from the correct road into brambles and marshlands. These errors are a social phenomenon with a social genesis.

This erroneous 'common understanding' can be remarkably prevalent. We sometimes find them in the finest universities; often, we find them in examination questions and sometimes they even mislead researchers.

Of course, maths and physics are not the only subjects which suffer under the weight of an erroneous 'common understanding'. All subjects suffer so. Even the English language is not exempt. Almost every day, the British Broadcasting Corporation, the BBC, receives letters from listeners complaining that newscasters or other broadcasters have made grammatical errors or have misused words. A daily example is the newscaster's use of the word condemn to mean contemn. Condemn means sentence to death; contemn means adversely criticise. Another example is a British Standard Assessment Test, SATs, exam paper for

11-year olds which asks, "What is the number on the dice?" while showing a picture of a single die. Dice is the plural of die.

We all make these errors. None of us can escape the whole of the erroneous 'common understanding'. We try to avoid errors as best we are able, we apologise for our slips, and we do not contemn too harshly the slips of others. Now we will look at a few examples of such errors.

Neutrino mass:

Durham University is a university of world renown in particle physics. I might be wrong, but I think the research papers published by the particle physics staff at Durham University are the most cited set of particle physics research papers of any university.

In the foyer of the physics department of Durham University, there is a poster which declares it is now known that neutrinos have mass. If you ask the particle physicists at Durham University, they will confirm that neutrinos have mass, and they will speak of many experiments that have detected neutrino oscillation across the three generations of neutrinos.

There is no evidence that neutrinos have mass; indeed, there is much evidence that the neutrino field is massless. It is remarkable that the physicists of an institution like Durham University should hold an erroneous 'common understanding' regarding something which might be considered the most central problem of particle physics. And, let us not blame only Durham; this misunderstanding is world-wide.

There is much evidence that the neutrino field is massless.

There is much evidence that the neutrino field squared is massive.

The neutrino field is massless:

Neutrinos have been shown to travel at the speed of light or very close to the speed of light in many experiments. In recent Lorentz violating frame experiments,[39] the margin of error in the ratio of velocity of the neutrino compared to the velocity of light has been reduced to $\frac{|v-c|}{c} < 5.6 \times 10^{-19}$. Neutrinos have been shown to be massless within very small margins of error in high end spectrum beta decay experiments for almost a century. There is much other evidence for neutrinos moving at the speed of light.

The neutrino field squared is massive:

Neutrinos come in three generations. These generations are the electron neutrino, v_e, the muon neutrino, v_μ, and the tau neutrino, v_τ. As neutrinos travel though our 4-dimensional space-time, they change from one type of a neutrino into a different type of neutrino. This has been observed many times. The probability of neutrinos changing type, oscillating is the technical phrase, is given by:

$$\text{Prob}(v_e \to v_\mu) = \sin^2(2\theta) \sin^2\left(\frac{(m_{v_e}^2 - m_{v_\mu}^2)}{4E} L\right) \quad (6.1)$$

We see that neutrinos can oscillate from one type to another if the neutrino field squared is massive; if the neutrino field squared was massless, the sine function would be zero and so the probability of oscillation would be zero.

Massive, massless, which is it?:

How can a field be massless and yet have mass when it is squared? Any type of complex number field can do this. Mass is a real number. Take the neutrino field to be a single complex number with zero real part. The square of that field is massive:

---

[39] Steckler, Floyd W. (2014) Constraining superluminal Electron and Neutrino Velocities using the 2010 Crab Nebula Flare and the Ice Cube PeV Neutrino Events : Astroparticle Physics 56 16018 arXiv 1306-6095.

$$\begin{bmatrix} 0 & b \\ -b & 0 \end{bmatrix} \begin{bmatrix} 0 & -b \\ b & 0 \end{bmatrix} = \begin{bmatrix} b^2 & 0 \\ 0 & b^2 \end{bmatrix} \quad (6.2)$$

In fact, we now believe the neutrino field is a massless quaternion field rather than a 2-dimensional complex field[40].

There is nothing in equation (6.1) that says the neutrino field is massive. The equation (6.1) says the neutrino field squared is massive.

We do not normally think of physicists as being careless. Why has all the evidence for massless neutrinos been ignored? There seems to be two reasons. The first of these reasons is that physicists are human; human beings make mistakes; it is what we do. The second of these reasons is an example of how the erroneous 'common understanding' can promulgate itself. The above equation, (6.1), is often presented as:

$$\text{Prob}(v_e \to v_\mu) = \sin^2(2\theta)\sin^2\left(\frac{\Delta m^2}{4E}L\right) \quad (6.3)$$

It is not clear that $\Delta m^2 = m_{v_e}^2 - m_{v_\mu}^2$. A little sloppy notation can go a long way.

Special relativity is 2-dimensional:
Many text books present special relativity as being a 4-dimensional theory. Special relativity is a 2-dimensional theory. Special relativity is simply the invariance of physics under the 2-dimensional space-time rotation matrix:

$$\begin{bmatrix} \cosh \chi & \sinh \chi \\ \sinh \chi & \cosh \chi \end{bmatrix} \quad (6.4)$$

Special relativity simply says that all physics is the same in all directions in 2-dimensional space-time. Different directions in 2-dimensional space-time are different velocities.

---

[40] See: Dennis Morris : The Quaternion Dirac Equation.

General relativity is a 4-dimensional theory. Special relativity is linear because it is concerned with a divisional algebra space (all division algebras are linear). General relativity is non-linear because it is concerned with a fabricated space rather than a division algebra space.

The erroneous 'common understanding' that special relativity is 4-dimensional has two sources. The first source is the idea that, since our universe is 4-dimensional, special relativity should be 4-dimensional. The primary individual responsible for this view is Hermann Minkowski (1864-1909) who rewrote special relativity in 1907 as a 4-dimensional theory[41]. Minkowski 4-dimensional space-time is the space-time of general relativity.

The mathematical expression of this 4-dimensionality is the use of 4-vectors to do calculations within special relativity. 4-vectors have a place in general relativity. Although unnecessarily complicating the calculations within special relativity, 4-vectors work most of the time, but they do not work all of the time. The norm of the 4-acceleration turns out to be imaginary, and so we have to sweep a minus sign under the carpet[42]. All that can be done in special relativity with 4-vectors can be more easily done with the matrix above, (6.4).

The second source of the 4-dimensional view of special relativity is the fact that it is intimately entangled with electromagnetism and electromagnetism is 4-dimensional. Electromagnetic calculations are the major use of 4-vectors.

There are no known dire consequences of taking special relativity to be 4-dimensional rather than 2-dimensional. However, it might be connected to the failure to find any super-symmetric particles.

Super-symmetry postulates one 2-dimensional Lorentz transformation of the form (6.4) and two types of particles. One of these types of particles is said to transform as a set of spin one particles under the 2-dimensional Lorentz transformation, (6.4). The other of these types of particles is said to transform as a set of spin one-half particles under the

---

[41] Minkowski Hermann (1908) Die Grundgleichungen fur die elektromagnetishen Vorgange in bewegten Korper.
[42] See : Dennis Morris : Empty Space is Amazing Stuff : pg 194.

2-dimensional Lorentz transformation, (6.4). We know there are two types of Lorentz transformations. There is the 2-dimensional Lorentz transformation, (6.4), which is spin one, and there is the 4-dimensional Lorentz transformation:

$$\begin{bmatrix} \cosh\chi & \sinh\chi & 0 & 0 \\ \sinh\chi & \cosh\chi & 0 & 0 \\ 0 & 0 & \cosh\chi & -\sinh\chi \\ 0 & 0 & -\sinh\chi & \cosh\chi \end{bmatrix} \quad (6.5)$$

This, (6.5), is a spin one-half Lorentz transformation. Perhaps super-symmetry can be rewritten with two Lorentz transformations and only one set of particles. At this time, this is mere speculation, but perhaps super-symmetry is being misled by the view that special relativity is 4-dimensional.

The teaching of special relativity:
In general, in the opinion of your author, special relativity is poorly taught within our universities. There is much talk of light beams and lightning bolts in this tuition. This is the way Einstein first deduced special relativity in 1905, and the teaching emulates Einstein's deduction.

By 1907, Minkowski had shown that special relativity is just rotation in 2-dimensional space-time, (6.4), although he presented this in four dimensions. In your author's opinion, Minkowski's presentation of special relativity is far better, in spite of it being 4-dimensional, than the light beams and lightning bolts presentation. Some courses on special relativity do not even mention Minkowski space-time.[43]

---

[43] The Minkowski presentation, in 2-dimensions, can be found in Dennis Morris : Empty Space is Amazing Stuff.

Careless Errors

The petty magnetic dipole error:
It is often said that a compass needle is attracted to the geographic north pole of the Earth. If this were so, there would be flocks of compass needles flying north every spring. The Earth's magnetic field is effectively uniform; the non-uniformities are so small as to be undetectable without specialist equipment. In a uniform magnetic field, a magnetic dipole (compass needle) aligns with the field. In a non-uniform magnetic field, we find attraction. The Earth's uniform magnetic field runs toward the geographic North Pole, and so a compass needle aligns to point to the Earth's geographic North Pole.

Upon one end of the compass needle is a little upper case N. Nowadays, this is often taken to be the north pole of the compass needle; it is actually the 'north seeking pole' of the compass needle as it was referred to in text-books of the 19th century – it seeks the north. If the magnetic field of the Earth is such that the north magnetic pole of the Earth coincides with the geographical north pole, then the end of the compass needle with the N on it is a magnetic south pole of the compass needle because 'like poles repel and unlike poles attract'; there's that 'attract' word again.

Actually, an evenly balanced compass needle dips a little to point through the Earth directly to the North Pole. You can show this by pinning the compass needle to a cork and floating it in a wine glass. The cork will not move to the northward side of the glass, but will align with the Earth's magnetic field and it will tilt. Exactly this experiment was done in 1581 by the mariner Robert Norman[44]. Compass needles are unevenly balanced within a manufactured compass to counteract this tendency to tilt. The tilting tendency varies with the latitude of the compass.

Conservation of energy properly presented:
The law of conservation of energy is normally presented as, "Energy can be neither created nor destroyed". If I were to measure the kinetic

---

[44] Norman Robert : The Newe Attractive – 1581 : see: ISBN 978-90-221-0616-7

energy of a cannon ball which was stationary relative to myself, I would get zero kinetic energy. If I were then to mount my bicycle and measure the kinetic energy of the same canon ball while I was moving at, say, 20 meters per second relative to the canon ball, I would not get zero. Lo! Behold! I am the superman. I have created energy.

If, say, one mega-joule of energy disappears from my room and simultaneously reappears on the moon, then energy has been neither created nor destroyed, but I am still missing a mega-joule of energy within my room.

The conservation of energy applies only within an inertial reference frame and energy is conserved locally. We have, "Energy can be neither created nor destroyed locally within an inertial reference frame".

The sloppy teaching of such basic physics must surely confuse and disenchant students with both economic and social implications.

Of course, the uncertainty principle allows the creation of miniscule amounts of energy for miniscule periods of time. Conservation of energy does not apply in the sub-atomic realm within miniscule periods of time. Conservation of energy does apply in the sub-atomic realm over any period of time greater than miniscule. The conservation of energy is a macroscopic law of physics.

Wave-particle duality:
Every student of physics will have been told that an electron is both a wave and a particle. If the student expresses disbelief, the student will be told that the disbelief is a failing of her intellect. Do we really believe that an electron is both a wave and a particle?

We opine that there is an error of understanding here. We offer an alternative clarification. This clarification is not universally accepted, and it might be in error itself. None-the-less, we are confident that there is some error of understanding in the wave-particle view of the electron.

The wave-particle duality of the electron is often expressed as electrons have both particle-like properties and wave-like properties. The particle properties are that the electron behaves as if it is a point of zero spatial extent in our 4-dimensional space-time. The wave properties are that an electron has non-zero spatial extent when it passes simultaneously through two spatially separated slits. Passing through two slits leads to interference patterns which are a hallmark of waves. Well! any single 'pulse' or object will interfere with itself if it passes simultaneously through two spatially separated slits.

What we mean by wave-particle duality is that sometimes the electron occupies zero space within our 4-dimensional space-time but it endures through time and at other times the electron occupies some space only instantaneously as it passes through two slits. We have a duality which is either 'no-space but some time' or 'no-time but some space'.

Waves and particles are a level of obfuscation covering the true nature of the electron's dual existence. Since spatial extent is associated with the part of the electron's dual existence which is usually called the wave-like properties of the electron and absence of spatial extent is associated with the part of the electron's existence which is usually called the particle-like properties of the electron, we might expect the distance function of space to be something to do with this observed duality of the electron.

It seems that the electron exists in the emergent quaternion space with distance function[45]:

$$d^2 = t^2 + x^2 + y^2 + z^2 \qquad (6.6)$$

This compares to the distance function of our 4-dimensional space-time which is:

$$d^2 = t^2 - x^2 - y^2 - z^2 \qquad (6.7)$$

We see that these two distance functions can coincide in only two cases. We have coincidence of distance functions if $x = y = z = 0$, in which case we have zero spatial extent which endures through time, and we have coincidence of distance functions if $t = 0$, in which case we have

---

[45] See: Dennis Morris : The Electron

spatial extent but only instantaneously. These matters are dealt with elsewhere in more detail[46].

The reader should be aware that the immediately above explanation of the dual nature of the electron is not universally known or accepted.

Summary:

Above, we have outlined some examples of the erroneous 'common understanding'. There are many more such examples. Some of these errors are quite petty like the compass needle being attracted to the North Pole, and some are easily rectified like the conservation of energy error. Others are quite profound like the wave-particle duality of the electron, and some, although profound, are surprisingly simple to rectify like the neutrino mass error.

These errors are not written into the mathematics of physics. Often these errors are errors of interpretation of the mathematics. Most of these errors are not profound errors, and they are not universally made by physicists. Many physicists will point out these errors to their students as they teach them the correct view. It is not that we have stumbled from the path. These errors are errors of sloppiness. None-the-less, it is remarkable how prevalent some of these errors are within the physics and mathematics community, and with that, these errors are an impediment to advancing our understanding.

---

[46] See : Dennis Morris : The Electron

Chapter 7

The All Rotations are 2-dimensional Error

Every rotation the reader, and all humankind, has ever seen is a 2-dimensional rotation. It is therefore understandable that until very recently mathematicians thought that all rotations are 2-dimensional. This is an error.

This error is not due to foolishness or laziness. It is an error of just not knowing any better. Neolithic farmers had around them all that was necessary to build a nuclear power station, but it never occurred to them to build a nuclear power station. Similarly, since group theory was discovered 150 year ago, mathematicians have had around them all that is necessary to form rotations of dimension other than two, but it did not occur to them to form a rotation of dimension other than two.

Higher dimensional rotations:
There are consequences beyond only complex numbers which flow from our finite group derivation of the 2-dimensional complex numbers.

Each of the higher dimensional complex numbers has a polar form like (4.14) & (4.18). The polar form of a division algebra is a radial variable and a set of angular variables. Using the matrix notation, the angular variables present themselves as a rotation matrix. Above, we have seen the rotation matrix of 2-dimensional space-time, (4.14), and the rotation matrix of 2-dimensional Euclidean space, (4.18). The polar form of a 3-dimensional complex number includes a 3×3 rotation matrix:

$$C_3H^2: \ \exp\left(\begin{bmatrix} 0 & b & c \\ c & 0 & b \\ b & c & 0 \end{bmatrix}\right) = \begin{bmatrix} v_A(b,c) & v_B(b,c) & v_C(b,c) \\ v_C(b,c) & v_A(b,c) & v_B(b,c) \\ v_B(b,c) & v_C(b,c) & v_A(b,c) \end{bmatrix} \quad (7.1)$$

This, (7.1), is an entirely new type of rotation; it is 3-dimensional rotation. This is something completely different from anything we have ever experienced. The functions within the 3-dimensional rotation matrix are a set of 3-dimensional trigonometric functions. Notice that they each have two arguments, two variables, rather than the one argument we find in the 2-dimensional trigonometric functions in (4.14) & (4.18).[47]

All of a sudden, we have expanded trigonometry into every dimension. The 2-dimensional Euclidean trigonometric functions were first discovered in India in the 6th century by Aryabhata (476-550). The 2-dimensional hyperbolic trigonometric functions, (4.14), were first used by Vincenzo Riccati (1707-1775) in his work *Opusculorum ad res physicas et mathmaticas pertinentium* (1757-1762) and later by Johann Heinrich Lambert (1728-1777)[48] in 1770. For centuries, mathematicians knew of only the 2-dimensional trigonometric functions. Our finite group derivation of the 2-dimensional complex numbers has led to a revolution in trigonometry; we now have trigonometric functions of every dimension.[49]

A revolution in trigonometry is a revolution in our understanding of rotation, and a revolution in our understanding of rotation is a revolution in our understanding of space.

Higher dimensional rotations again:
With the higher dimensional trigonometric functions, we get the higher dimensional rotations. One of the great errors of modern physics and of modern mathematics is the belief that the only rotations are the 2-dimensional rotations expressed as:

$$\begin{bmatrix} \cosh \chi & \sinh \chi \\ \sinh \chi & \cosh \beta \end{bmatrix} \& \begin{bmatrix} \cos \theta & \sin \theta \\ -\sin \theta & \cos \theta \end{bmatrix} \quad (7.2)$$

---

[47] Much more detail of the 3-dimensional trigonometric functions is given in : Dennis Morris : Complex Numbers The Higher Dimensional Forms.
[48] Johann Lambert was the first person to prove that pi is irrational.
[49] This is still an unexplored area of mathematics.

Above, (7.1), we have an example of a 3-dimensional rotation. This rotation does not have 2-dimensional sub-rotations because the finite group C_3 which underlies this rotation does not have a C_2 subgroup. The 3-dimensional rotation, (7.1), is not a composition of 2-dimensional rotations.

Within our 4-dimensional space-time, we find only (six off) 2-dimensional rotations. These 2-dimensional rotations can be composed together to form a 2-dimensional rotation in another 2-dimensional plane. The 3-dimensional rotation above, (7.1) is a rotation entirely different from a composition of 2-dimensional rotations; it is in a different type of space for a start.

We similarly have higher dimensional rotations derived from every finite group, and so we have 4-dimensional rotations, 5-dimensional rotations etc..

The 3-dimensional rotation above, (7.1), does not seem to play any part in physics, and this is also true of almost all higher dimensional rotations, but there is a one order four finite group, $C_2 \times C_2$, whose 4-dimensional rotations do play a role in physics[50]. We have the 4-dimensional (left-chiral) quaternion rotation matrix:

$$\mathbb{H}_{L\chi}^{Rot} = \begin{bmatrix} \cos\lambda & \frac{b}{\lambda}\sin\lambda & \frac{c}{\lambda}\sin\lambda & \frac{d}{\lambda}\sin\lambda \\ -\frac{b}{\lambda}\sin\lambda & \cos\lambda & -\frac{d}{\lambda}\sin\lambda & \frac{c}{\lambda}\sin\lambda \\ -\frac{c}{\lambda}\sin\lambda & \frac{d}{\lambda}\sin\lambda & \cos\lambda & -\frac{b}{\lambda}\sin\lambda \\ -\frac{d}{\lambda}\sin\lambda & -\frac{c}{\lambda}\sin\lambda & \frac{b}{\lambda}\sin\lambda & \cos\lambda \end{bmatrix} \quad (7.3)$$

$$\lambda = \sqrt{b^2 + c^2 + d^2}$$

Because the finite group $C_2 \times C_2$ has three C_2 sub-groups, this rotation matrix will reduce to three sub-rotations, but these sub-rotations are not

---

[50] See : Dennis Morris : Upon General Relativity and Dennis Morris : The Electron.

2-dimensional rotations; they are still 4-dimensional rotations. We can change the co-ordinate system by setting $c = d = 0$. This gives:

$$\mathbb{H}^{Rot}_{L\chi}\big|_{c=d=0} = \begin{bmatrix} \cos b & \sin b & 0 & 0 \\ -\sin b & \cos b & 0 & 0 \\ 0 & 0 & \cos b & -\sin b \\ 0 & 0 & \sin b & \cos b \end{bmatrix} \quad (7.4)$$

If we take the eigenvectors of the single variable quaternion rotation matrix, (7.4), we do not get two eigenvectors which are independent of the angle of rotation. All eigenvectors of this rotation matrix depend upon the angle of rotation. This is not 2-dimensional rotation about an axis (or two).

Quaternion rotation, (7.4), is a double cover rotation. The top left-hand corner is a clockwise-rotation, and the bottom right-hand corner is an anti-clockwise rotation. A rotation in quaternion space is a rotation both clockwise and anti-clockwise at the same time. To rotate all the way around back to where we started requires 720^0; that is 360^0 in the clockwise direction and 360^0 in the anti-clockwise direction. Quaternion space is believed to be the space of the electron and the neutrino.

Summary:

Above, we have given just two examples of higher dimensional rotations. Of course each type of rotation is associated with a kind of rotational symmetry. Each type of rotation is also associated with a kind of reflective symmetry in that the trigonometric functions must necessarily exhibit a kind of 'oddness' and 'evenness'. Within the higher dimensional rotations shown above, we have discovered 'double cover' rotation and absence of 2-dimensional sub-rotation. There are four-fold cover rotations in the 8-dimensional $C_2 \times C_2 \times C_2$ spaces and eight-fold cover rotations in the 16-dimensional $C_2 \times C_2 \times C_2 \times C_2$ spaces etc..

The All Rotations are 2-dimensional Error

The higher dimensional rotations 'fold' into lower dimensional rotations. They have to do this because the higher order groups reduce to the lower order groups when some of their variables are set to zero and so the higher order spaces must reduce to the lower order spaces when some of the variables are zero[51]. Of course, setting variables to zero is just rotating the co-ordinate system.

Our concept of rotation, and of the nature of empty space, has expanded considerably.

Of course, if it had occurred to our Neolithic ancestors to build a nuclear power station, their concept of reality would have similarly expanded considerably.

---

[51] See : Dennis Morris : The Naked Spinor

Chapter 8

Ɵersted and the Breakdown of Parity Error

We now present a lighter chapter with an historical narrative.

The discovery of electromagnetism:
Although magnetism has been studied seriously since Petrus Peregrinus de Maricourt (1240-unknown) wrote his book 'The Letter of Petrus Peregrinus on the Magnet' in 1269 and William Gilbert (1544-1603) published 'De Magnete' in 1600, the association of magnetism with electric current dates from only just later than 1800.

On 21st April 1820, the Danish physicist Hans Christian Ɵersted (1777-1851) was delivering a physics lecture in which he intended to demonstrate that an electric current flowing through a wire did not deflect a compass needle. He noticed that when he switched the current on and off, the compass needle did move. Even more surprising to Ɵersted, the compass needle was deflected by a force perpendicular to the direction of the current across the wire rather than away from or towards the wire. We've all had days like this.

Previous to this observation by Ɵersted, all physicists had assumed that any force, magnetic or otherwise, caused by the electric current in the wire would be parallel to the wire. As had others before him, Ɵersted had sought such parallel forces and not found them. Neither Ɵersted nor anyone else had imagined that a current moving in a given direction could produce a force in any direction other than parallel to the current. The magnetic force detected by Ɵersted pointed in a circle around the wire as his later researches confirmed.

Just think about this for a while. Force fields like gravity or the electric field 'point' in a straight direction directly towards the object, charge, 'causing' the force. A force that 'points' in a circle is something completely different. There is no charge associated with such a force; there is no object to 'cause' the force. However, if a current in a wire is going to 'cause' a force perpendicular to the direction of the current across the wire rather than away from or towards the wire, then, from symmetry considerations, that force must 'point' in a circle.

In picture form[52]:

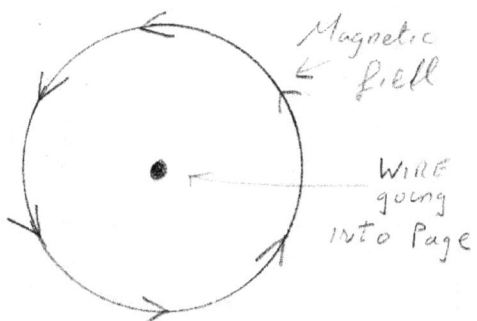

Prior to Ꝋersted's investigations, physicists had unwittingly made an assumption regarding the direction of forces associated with currents in wires which turned out to be in error. The physicists made an error, but it was an error they soon remedied. The reader might be wondering why your author bothered reporting an error of such little import.

Historical note: A similar discovery of the rotational nature of the magnetic field generated by an electric current was reported in the 3rd August 1802 edition of *Gazetta di Trentino*. The discovery was made and reported by the Italian physicist Gian Domenico Romagnosi (1761-1835). Romagnosi discovered this rotational nature of the magnetic field, but Ꝋersted gets the credit.

Today, the magnetic field is measured in the SI unit which is the Tesla. The older CGS unit measuring the magnetic field was the Gauss. A hundred years ago, the magnetic field was measured in amps per metre. The amps per metre unit was called the Ꝋersted.

---

[52] I might have got the arrows pointing in the wrong direction. The conventional direction of the magnetic field is arbitrary.

The year 1820 also saw Andre-Marie Ampere (1775-1836) give a mathematical form to Ɵersted's discovery and Jean-Baptiste Biot (1774-1862) and Felix Savart (1791-1841) propose the Biot-Savart law of electromagnetism. Only eleven years later, on 28th October 1831 Michael Faraday (1791-1867), building on the work of Humphey Davy (1778-1829) and William Hyde Wollaston (1766-1828) showed that a changing magnetic field produces an electric current.

The magnetic field is left-handed:
We have reported the minor error above because of what happened next. Physicists saw into the heart of Ɵersted's observation, and, instead of making an error of omission, they got things bang right. There is an imbalance in the magnetic field; it rotates in only one direction; it is left-handed. The physicists did not omit to see this. This means that the universe does not treat right-hands and left-hands equally.

Instead of speaking of current, let us speak of a stream of electrons moving along the wire. As the electrons move along the wire, the magnetic field rotates in only one direction; it does not rotate in a balanced way[53]. It is true that if the electrons move in the opposite direction then the magnetic field rotates in the opposite direction, but this is only a change of the perspective of the observer. The imbalance is still there. This is violation of the left/right balance of the universe. Today, it would be called a violation of parity.

Perhaps the reader is unsure of this imbalance; it is not that clear. The easiest way to see that the magnetic field is left-handed is to look at the electromagnetic tensor.

First we present the two types of quaternions, left-chiral and right-chiral:

---

[53] Rotation in quaternion space is both clockwise and anti-clockwise at the same time. If the rotational nature of the magnetic field had been like quaternion rotation, there would have been a balance within the universe. We cannot observe quaternion rotation in our 4-dimensional space-time.

$$\mathbb{H}_{L\chi} = \begin{bmatrix} a & b & c & d \\ -b & a & -d & c \\ -c & d & a & -b \\ -d & -c & b & a \end{bmatrix}, \quad \mathbb{H}_{R\chi} = \begin{bmatrix} a & b & c & d \\ -b & a & d & -c \\ -c & -d & a & b \\ -d & c & -b & a \end{bmatrix} \quad (8.1)$$

$$\hat{i}\hat{j} = k, \hat{j}\hat{i} = -k, \qquad \hat{i}\hat{j} = -k, \hat{j}\hat{i} = k$$

The commutation relations between the imaginary units, $\{\hat{i}, j, k\}$, are opposite in the two types of quaternions, (8.1). We see that the difference between these two types of quaternions is in the distribution of the minus signs in the 3×3 part in the bottom right-hand corner of the matrices. It has been shown elsewhere[54] that the classical electromagnetic tensor emerges from the superimposition of the four differentials, two E-fields and two B-fields, of the two quaternion potentials. That classical electromagnetic tensor is:

$$\text{Emag tensor} = \begin{bmatrix} 0 & E_x & E_y & E_z \\ -E_x & 0 & -B_z & B_y \\ -E_y & B_z & 0 & -B_x \\ -E_z & -B_y & B_x & 0 \end{bmatrix} \quad (8.2)$$

We see that the electric field, E_i, matches the distribution of minus signs in both the left-chiral and the right-chiral types of quaternions but that the magnetic field, B_i, matches the distribution of minus signs in only the left-chiral quaternion. The magnetic field is left-handed.

The imbalance was spotted by physicists contemporary with Ørsted and frequently referred to by them – well spotted. This imbalance remained a hot topic of conversation until the positron was predicted in 1928 and was discovered in 1932. If we use positrons instead of electrons, the magnetic field is reversed in direction. It seems as though the universe has been rebalanced until we realise that there is a lot less anti-matter in the universe than matter. Ours is a matter universe; thus,

---

[54] See : Dennis Morris : The Physics of Empty Space.

the imbalance persists; the magnetic field is left-handed; the universe still violates parity.

In 1932, following the discovery of the positron, all talk of the imbalance of the magnetic field went away. Having initially seen things correctly, physicists forgot the correct understanding. By the early 1950's, violation of parity, that which had been a hot topic of discourse for over a century prior to 1928, was completely forgotten. This is a quite amazing event in our saga of errors. A significant aspect of theoretical physics had been understood and was then forgotten.

In the latter part of 1956, Madam Chien-Shiung Wu (1912-1997)[55] showed that parity is violated by the weak force as the physicists Tsung-Dao Lee (1926-) and Chen-Ning Franklin Yang (1922-) had conjectured earlier in 1956[56]. We now know that the neutrino, which is associated with the weak force, does violate parity. The neutrino is left-handed; the anti-neutrino is right-handed.

The neutrino field comes out of the relativistic energy momentum relation as the massless superimposition of the B-fields of the two quaternion non-commutative differential operators. The superposition of the two B-fields from the two quaternions, left-chiral and right-chiral, is a left-chiral field.[57]

Interim summary:
Physicists had found violation of parity in 1820; they were aware of it until 1932; then they forgot about it and had to rediscover it. It is not only that mathematicians and physicists make errors as they struggle to understand the universe; it is also that they unmake non-errors.

---

[55] Wu C. S. Ambler E. Hayward R. W. Hoppes D. D. Hudson R. P. (1957) Experimental Test of Parity Conservation in Beta Decay Physical Review 105 (4) 1413-1415

[56] Lee T. D. Yang C. N. 1st October 1956 Question of Parity Conservation in Weak Interactions : Physical Review 104 (1) 254

[57] See : Dennis Morris : The Quaternion Dirac Equation

The association between the magnetic field and the neutrino is only just beginning to be understood.

Missed special relativity:
With Ørsted's discovery in 1820 of the magnetic force generated by an electric current, physicists were very close to discovering special relativity. Indeed, it was this very phenomenon which led Einstein to discover special relativity 85 years later. All we need is the idea that the electric current is some form of matter moving within the wire, a stream of electrons, and a person on a bicycle travelling alongside the wire at the same speed as the matter moving within the wire. Of course, there was no concept of electrons in 1820, and perhaps 85 years is not too long to wait. Physicists in the 19th century did not miss special relativity because they took the wrong path; they missed special relativity because they did not put their best foot forward.

The nature of our space-time:
There is something else that Ørsted's discovery of 1820 points towards but that physicists of both the 19th century and the 20th century have missed. If our 4-dimensional space-time is no more than four independent copies of the real numbers, $\mathbb{R}^4$, then how can a current moving in one of these $\mathbb{R}$ directions cause a force in a different independent $\mathbb{R}$ direction?

There are no mathematical relations between two copies of the real numbers. Each copy of the real numbers is a separate and independent closed division algebra. Only spinor spaces like the complex plane have mathematical connections between the orthogonal directions within these spaces. These connections are things like $\hat{i}^2 = -1$ in the 2-dimensional complex numbers, $\mathbb{C}$, or $\hat{p}^3 = +1$ in the 3-dimensional complex numbers.

That a current moving in one direction can cause a force in an orthogonal direction clearly indicates that we live in a space-time that is something more than only four copies of the real numbers. There must be some kind of spinor space structure within our space-time. The only spinor spaces which have anything like right-angles in them are the 2-dimensional complex numbers, $\mathbb{C}$, (genuine right-angle) and the 4-dimensional quaternions (double cover right-angle) and some of the higher dimensional $C_2 \times C_2 \times ...$ algebras (quadruple cover ...).

We have a clear indication that the nature of our 4-dimensional space-time is more than only $\mathbb{R}^4$. We have had this indication for nearly 200 years, and yet modern physics, including string theory, is formulated in $\mathbb{R}^n$ space. We are looking but not seeing.

Chapter 9

The Rotation about an Axis Error

It is amazing how you can look at something every day of your life and not see it. An example of this is our familiar 2-dimensional rotation in the Euclidean plane. Everybody knows that rotation is about an axis. The Earth spins on an axis; tops spin on an axis; for decades, it has been assumed that electrons spin on an axis.

There is rotation in the complex plane, $\mathbb{C}$, expressed as the rotation matrix:

$$\begin{bmatrix} \cos b & \sin b \\ -\sin b & \cos b \end{bmatrix} \qquad (9.1)$$

This is a 2-dimensional rotation. The complex plane, $\mathbb{C}$, is a 2-dimensional space. This, (9.1), is not rotation about an axis. There is no third dimension sticking out of the complex plane to be an axis of rotation – the complex plane is 2-dimensional. Rotation in a spinor space, such as the complex plane, $\mathbb{C}$, is not rotation about an axis. The same is true of rotation in quaternion space or in the 3-dimensional spinor space above, (7.1).

Let us do the mathematics. An axis of rotation is a direction in space which is unaffected by the rotation; this is a direction which is independent of the angle of rotation. Let us take the eigenvectors of (9.1). There are only two of them because this is a 2-dimensional space; these two are:

$$\begin{bmatrix} 0 \\ e^{ib} \end{bmatrix} \& \begin{bmatrix} e^{-ib} \\ 0 \end{bmatrix} \qquad (9.2)$$

We have no direction in this space which is independent of the angle of rotation; hence, this is not rotation about an axis.

With a little thought, we realise that 2-dimensional rotation in our 4-dimensional space-time must be rotation about two axes because there are two axes which are independent of the angle of rotation. A rotation matrix in our 4-dimensional space-time is of a form similar to:

$$\text{4-dim Space-time} = \begin{bmatrix} \cosh\chi & \sinh\chi & 0 & 0 \\ \sinh\chi & \cosh\chi & 0 & 0 \\ 0 & 0 & 1 & 0 \\ 0 & 0 & 0 & 1 \end{bmatrix} \quad (9.3)$$

If we take the four eigenvectors of this matrix, we get two eigenvectors which are independent of the angle of rotation:

$$\begin{bmatrix} 0 \\ 0 \\ 1 \\ 0 \end{bmatrix} \& \begin{bmatrix} 0 \\ 0 \\ 0 \\ 1 \end{bmatrix} \quad (9.4)$$

Rotation in our 4-dimensional space-time is 2-dimensional. This leaves two directions unaffected by the rotation. Rotation in our 4-dimensional space-time is rotation about two axes.

Of course, there is a 3-dimensional spatial sub-space within our 4-dimensional space-time. Rotation within this spatial sub-space is expressed in a rotation matrix similar to:

$$\begin{bmatrix} \cos\theta & \sin\theta & 0 \\ -\sin\theta & \cos\theta & 0 \\ 0 & 0 & 1 \end{bmatrix} \quad (9.5)$$

The three eigenvectors of this matrix include one eigenvector which is independent of the angle of rotation:

$$\begin{bmatrix} 0 \\ 0 \\ 1 \end{bmatrix} \quad (9.6)$$

The Rotation about an Axis Error

This, (9.5), is 2-dimensional rotation in a 3-dimensional sub-space. There is one direction spare to be an axis.

For physicists:
For decades, physicists have struggled to understand electron spin[58]. For decades, it has been assumed that electron spin is a 2-dimensional rotation about a single axis. With hindsight, this seems very foolish, but with no other kind of rotation known, how could they have done differently?

An electron has been described by a pair of complex numbers since the 1920's in both the Pauli modification of the Schrödinger equation (1927)[59] and in the Dirac equation (1928)[60]. A pair of complex numbers is just an obscure way of writing a quaternion.

Within the 3-dimensional spatial sub-space of our 4-dimensional space-time, all rotations are 2-dimensional and about a single axis, but within quaternion space, electrons do not rotate about an axis. They also have twice the rotation we would expect and so twice the magnetic field we would expect – see (7.3) & (7.4).

Of course, an observer in our 4-dimensional space-time is able to observe only 2-dimensional rotation, and 2-dimensional spatial rotation is about a single axis in the 3-dimensional spatial sub-space of our 4-dimensional space-time.

No rotation in the complex plane:
The astute reader will recall that we above considered a plate spinning horizontally on top of a rod in a circus act. The astute reader will have

---

[58] See : Dennis Morris : The Electron : for a more detailed exposition of electron spin.
[59] Wolfgang Pauli : (1927) Zur Quantenmechanik des magnetishen Elektrons Zeitschrift fur Physik (43) 601-623
[60] Dirac P. A. M. (1928) The Quantum Theory of the Electron. Proceedings of the Royal Society A: Mathematical, Physical and Engineering Sciences 117 (778)

noticed that at the start of this chapter, circa (9.1), we referred to rotation in the complex plane as if it really exists.

Thinking back to the spinning plate, we realised that the spinning plate is rotating in three 2-dimensional planes; these are one Euclidean plane and two hyperbolic space-time planes because any point on the edge of the plate must accelerate and decelerate (that's rotate in 2-dimensional space-time) in two orthogonal spatial directions.

Can we rotate in the complex plane, $\mathbb{C}$? There is no space-time in the complex plane. There is no rotation in the complex plane, but there is a rotation matrix.

The uniqueness of our 4-dimensional space-time:
Since, as shown by the spinning plate, we need three 2-dimensional spaces, two of a space-time nature and one of a Euclidean nature, to allow rotation in a Euclidean plane, we cannot have Euclidean rotation unless we have a space which contains at least these three 2-dimensional spaces. Such a space must have a distance function which respects both types of 2-dimensional rotation because rotation holds distance from the centre of rotation invariant. This means that only distance functions which are quadratic forms can hold a 2-dimensional Euclidean rotation.

Neither of the two 2-dimensional spinor spaces (division algebra spaces) can support both types of 2-dimensional rotation. Thus any space which has a spinning plate within it must be of a dimension higher than two and have a distance function which is a quadratic form.

There is a similar constraint on 3-dimensional rotations. There are three types of 3-dimensional rotations. These spaces must be mixed together to allow any type of 3-dimensional rotation.

It is a simple algebraic fact of group theory and of the algebras that finite groups contain that only the finite group $C_2 \times C_2$ contains spaces that allow any kind of physical rotation. The unique space that allows this is our 4-dimensional space-time.

The Rotation about an Axis Error

Genesis of the spin axis error:

How did mathematicians and physicists come to think that all rotation is rotation about an axis when the complex plane is clearly 2-dimensional with no spare dimension to be an axis? We can ask a second question; how did mathematicians and physicists come to think that all rotation is rotation about one axis when it has been accepted for over a century that we live in a 4-dimensional space-time? There are clearly two spare dimensions in our 4-dimensional space-time to be axes.

Prior to Einstein publishing the special theory of relativity in 1905, it was generally accepted that our space-time was two separate spaces; these were a 1-dimensional temporal space and a 3-dimensional spatial space. It made numerical sense to see a 2-dimensional spatial rotation as having a single rotational axis in the 3-dimensional spatial space. When, in 1905, Einstein showed that our space-time is a single 4-dimensional space, the implications for a single rotational axis were simply not realised. Even so, how can you look at the 2-dimensional complex plane every day of your life and not realise that there is no third dimension to be an axis of rotation? It is a case of looking but not seeing.

We must conclude that mathematicians and physicists made this mistake because they unquestioningly accepted what had been the common mantra prior to 1905. Actually, many did not accept Einstein's special relativity at all in 1905. It seems that mathematicians and physicists are only human after all.

Even Mathematicians and Physicists make Mistakes

Chapter 10

The Complex Numbers Differentiation Error

Differentiation is forming a ratio of one variable to another variable. A differential is therefore a product of the form:

$$dy \frac{1}{dx} = \frac{dy}{dx} \qquad (10.1)$$

Products properly exist within only division algebras. One of the defining properties of a product is that of multiplicative closure. The product of any two elements of a division algebra must itself be an element of that division algebra. The product of two complex numbers must be a complex number, and so the differential of a complex function must be a complex function.

Differentiation within the complex numbers:
Clearly, a function of a complex variable is a function of two variables. For example:

$$\begin{aligned} f = z^3 &= (x+iy)^3 \\ &= (x^3 - 3xy^2) + i(3x^2y - y^3) \\ &= u(x,y) + iv(x,y) \end{aligned} \qquad (10.2)$$

A complex variable is two variables; one of these is a real variable, and the other of these is an imaginary variable. We therefore have four partial derivatives of (10.2):

$$\frac{\partial u}{\partial x} = 3x^2 - 3y^2 \qquad \frac{\partial u}{i \partial y} = \frac{1}{i}(-6xy) = i6xy$$

$$\frac{i \partial v}{\partial x} = i6xy \qquad \frac{i \partial v}{i \partial y} = 3x^2 - 3y^2 \qquad (10.3)$$

We have four derivatives, but a complex number is 2-dimensional; it has only two bits; it makes no sense to discard two of the four derivatives; what are we to do with the 'extra' derivatives?

Differentiating (forming a ratio of) a real function by a real variable gives a real function; we know that from the calculus of the real numbers. Similarly, differentiating an imaginary function by an imaginary variable will be a real function – the i s cancel. Similarly, the derivative of an imaginary function by a real variable will be an imaginary function and the derivative of a real function by an imaginary variable will be the negative of an imaginary function (the inverse of i is $-i$). Looking back at (10.3), by simplifying the $\dfrac{\partial u}{i\partial y}$ term, we see that we have only two complex functions. The conventional mantra is that a complex function is differentiable if and only if:

$$\frac{\partial u}{\partial x} = \frac{\partial v}{\partial y} \quad \& \quad \frac{\partial v}{\partial x} = -\frac{\partial u}{\partial y} \qquad (10.4)$$

These are the Cauchy-Riemann equations. The Cauchy Riemann equations cut four differentials down to two differentials and thereby allow us to form a complex number. Conventionally, if the Cauchy-Riemann equations are not satisfied by a complex function, then that function is not differentiable. Most complex functions do not satisfy the Cauchy-Riemann equations, and so most complex functions are not conventionally differentiable.

Integration:
We point out that in conventional complex analysis, a function is conventionally integrable if and only if it is Cauchy-Riemann differentiable.

Unsatisfactory differentiation:
Within the complex numbers, there are many functions which are continuous and which at first sight we would expect to be differentiable.

It seems utterly bizarre that only very 'special' functions which satisfy the Cauchy-Riemann equations should be differentiable. This is not what we might reasonably expect within a division algebra; it is certainly not what we find within the real numbers.

Differentiation within the complex numbers, $\mathbb{C}$ - a different view:
The reader might be tempted to form the complex differential as a ratio of two complex variables; that is, after all, the nature of a differential. Using common sense, we will expect get something like $\dfrac{d(a+ib)}{d(x+iy)}$ within which $a(x,y)$ & $b(x,y)$ are both functions of $\{x,y\}$. Let us evaluate this using the complex number multiplication:

$$\frac{da+idb}{dx+idy} = \frac{(da+idb)(dx-idy)}{dx^2+dy^2}$$

$$= \frac{dadx+dbdy+i(-dady+dbdx)}{dx^2+dy^2} \qquad (10.5)$$

$$= \frac{da}{dx+\frac{dy^2}{dx}} + \frac{db}{\frac{dx^2}{dy}+dy} - i\frac{da}{\frac{dx^2}{dy}+dy} + i\frac{db}{dx+\frac{dy^2}{dx}}$$

Ignoring the $\{dx^2, dy^2\}$ terms, we get:

$$\frac{d(a+ib)}{d(x+iy)} = \frac{da}{dx}+\frac{db}{dy}+i\left(\frac{db}{dx}-\frac{da}{dy}\right) \qquad (10.6)$$

We see that we have reduced the four differentials $\left\{\dfrac{\partial a}{\partial x},\dfrac{\partial a}{\partial y},\dfrac{\partial b}{\partial x},\dfrac{\partial b}{\partial y}\right\}$ to a real bit and an imaginary bit.

In matrix form:

$$\frac{d(a+ib)}{d(x+iy)} = \begin{bmatrix} \dfrac{da}{dx}+\dfrac{db}{dy} & \dfrac{db}{dx}-\dfrac{da}{dy} \\ -\left(\dfrac{db}{dx}-\dfrac{da}{dy}\right) & \dfrac{da}{dx}+\dfrac{db}{dy} \end{bmatrix} \qquad (10.7)$$

This is very different from the Cauchy-Riemann differentiation above, (10.4). What we have done (10.5) to (10.7) is quite simply differentiation from first principles within the complex numbers, but it effectively undermines a whole well established area of mathematics based upon the conventional Cauchy-Riemann way of differentiating within the complex numbers.

Matrix differentiation:
A much easier way of differentiating the complex numbers is to use the matrix form of the complex numbers. Since the multiplication operation of every division algebra is based on matrix multiplication, this technique can be applied to every division algebra. For higher dimensional division algebras, this technique is cumbersome involving many large matrices, but it is simple and clear.

We will differentiate the complex function[61]:

$$\begin{bmatrix} f(x,y) & g(x,y) \\ -g(x,y) & f(x,y) \end{bmatrix} \qquad (10.8)$$

With respect to the complex variable:

$$\begin{bmatrix} x & y \\ -y & x \end{bmatrix} \qquad (10.9)$$

We have:

---

[61] This is reproduced in Dennis Morris : Elementary Calculus from an Advanced Standpoint.

$$\frac{\partial\begin{bmatrix} f(x,y) & g(x,y) \\ -g(x,y) & f(x,y) \end{bmatrix}}{\partial \begin{bmatrix} x & y \\ -y & x \end{bmatrix}} \qquad (10.10)$$

$$= \frac{\partial\begin{bmatrix} f(x,y) & 0 \\ 0 & f(x,y) \end{bmatrix}}{\partial\begin{bmatrix} x & 0 \\ 0 & x \end{bmatrix}} + \begin{bmatrix} 0 & 1 \\ -1 & 0 \end{bmatrix}\frac{\partial\begin{bmatrix} g(x,y) & 0 \\ 0 & g(x,y) \end{bmatrix}}{\partial\begin{bmatrix} x & 0 \\ 0 & x \end{bmatrix}}$$

$$+ \frac{\partial\begin{bmatrix} f(x,y) & 0 \\ 0 & f(x,y) \end{bmatrix}}{\begin{bmatrix} 0 & 1 \\ -1 & 0 \end{bmatrix}\partial\begin{bmatrix} y & 0 \\ 0 & y \end{bmatrix}} + \begin{bmatrix} 0 & 1 \\ -1 & 0 \end{bmatrix}\frac{\partial\begin{bmatrix} g(x,y) & 0 \\ 0 & g(x,y) \end{bmatrix}}{\begin{bmatrix} 0 & 1 \\ -1 & 0 \end{bmatrix}\partial\begin{bmatrix} y & 0 \\ 0 & y \end{bmatrix}}$$

(10.11)

We have reduced everything to real number differentiation which we understand. Of course, we have four differentials within (10.11). Tidying this, (10.11), gives:

$$\frac{\partial\begin{bmatrix} f(x,y) & g(x,y) \\ -g(x,y) & f(x,y) \end{bmatrix}}{\partial\begin{bmatrix} x & y \\ -y & x \end{bmatrix}} = \begin{bmatrix} \frac{\partial f}{\partial x} + \frac{\partial g}{\partial y} & \frac{\partial g}{\partial x} - \frac{\partial f}{\partial y} \\ -\left(\frac{\partial g}{\partial x} - \frac{\partial f}{\partial y}\right) & \frac{\partial f}{\partial x} + \frac{\partial g}{\partial y} \end{bmatrix} \qquad (10.12)$$

$$= \begin{bmatrix} Div & Curl \\ -Curl & Div \end{bmatrix}$$

This is the same as (10.7) above.

Other than the fact that it is obviously correct, there is a clear advantage to this non-conventional matrix differentiation, (10.12). Every continuous complex function is differentiable with this matrix differentiation. Using Cauchy-Riemann differentiation, we can differentiate only a very small proportion of complex functions; matrix differentiation allows us to differentiate any reasonable complex function. Furthermore, matrix differentiation, appropriately adjusted, will apply within any division algebra; it is not restricted to the 2-dimensional complex numbers.

A complex potential:

We see that the differential of a complex function is a divergence and a curl, (10.12). What kind of thing is it whose differential is a divergence and a curl? It is a potential like the potential that leads to an electric field and a magnetic field by differentiation.

So, if we have a complex function over two of our space-time dimensions, we have a potential over two of our space-time dimensions. Take three such complex functions set at right-angles to each other, and we have a 3-dimensional vector potential.

People familiar with gauge covariant differentiation, and $U(1)$ gauge covariant differential will realise how neatly the idea of a locally varying complex function being a potential fits in with QED.

Setting $g(x,y)=0$ in (10.12) gives the gradient:

$$\frac{\partial \begin{bmatrix} f(x,y) & 0 \\ 0 & f(x,y) \end{bmatrix}}{\partial \begin{bmatrix} x & y \\ -y & x \end{bmatrix}} = \begin{bmatrix} \frac{\partial f}{\partial x} & -\frac{\partial f}{\partial y} \\ \frac{\partial f}{\partial y} & \frac{\partial f}{\partial x} \end{bmatrix} = grad(f) \quad (10.13)$$

We see that the gradient is a 1-form (conjugate).

Of course, the above assumes that all the derivatives exist at the points of evaluation.

Differential operators:

Looking at (10.12), we notice that the differential is of the form of a matrix product:

$$\begin{bmatrix} \partial x & -\partial y \\ \partial y & \partial x \end{bmatrix} \begin{bmatrix} f(x,y) & g(x,y) \\ -g(x,y) & f(x,y) \end{bmatrix} \quad (10.14)$$

Because differentiation is a linear operation and matrix multiplication is linear multiplication, we can 'invent' a calculative shortcut to assist in differentiating:

$$\frac{\partial \begin{bmatrix} f(x,y) & g(x,y) \\ -g(x,y) & f(x,y) \end{bmatrix}}{\partial \begin{bmatrix} x & y \\ -y & x \end{bmatrix}} \equiv \begin{bmatrix} \partial x & -\partial y \\ \partial y & \partial x \end{bmatrix} \begin{bmatrix} f & g \\ -g & f \end{bmatrix} \quad (10.15)$$

$$= \begin{bmatrix} \dfrac{\partial f}{\partial x} + \dfrac{\partial g}{\partial y} & \dfrac{\partial g}{\partial x} - \dfrac{\partial f}{\partial y} \\ -\left(\dfrac{\partial g}{\partial x} - \dfrac{\partial f}{\partial y}\right) & \dfrac{\partial f}{\partial x} + \dfrac{\partial g}{\partial y} \end{bmatrix} \quad (10.16)$$

We call the matrix:

$$d = \begin{bmatrix} \partial x & -\partial y \\ \partial y & \partial x \end{bmatrix} \quad (10.17)$$

the differentiation operator. Notice the position of the minus sign in the differentiation operator (10.17). The differential operator is the inverse of the matrix form of an algebra.

The differentiation operator acts as if by matrix multiplication.

Because the complex numbers are commutative, this operator can act from either the right or from the left with the same results:

$$\begin{bmatrix} f & g \\ -g & f \end{bmatrix} \begin{bmatrix} \partial x & -\partial y \\ \partial y & \partial x \end{bmatrix} = \begin{bmatrix} \dfrac{\partial f}{\partial x} + \dfrac{\partial g}{\partial y} & \dfrac{\partial g}{\partial x} - \dfrac{\partial f}{\partial y} \\ -\left(\dfrac{\partial g}{\partial x} - \dfrac{\partial f}{\partial y}\right) & \dfrac{\partial f}{\partial x} + \dfrac{\partial g}{\partial y} \end{bmatrix} \quad (10.18)$$

We can separate this differentiation operator into its real part and its imaginary part:

$$\begin{bmatrix} \partial x & 0 \\ 0 & \partial x \end{bmatrix} \equiv \frac{\partial}{\partial x}, \quad \begin{bmatrix} 0 & -\partial y \\ \partial y & 0 \end{bmatrix} \equiv -i\frac{\partial}{\partial y} = p_x \quad (10.19)$$

We see the imaginary part is what is called the momentum operator in quantum mechanics. We see that the momentum operator is no more than 'differentiate with respect to minus the imaginary variable'. The energy operator is then differentiate with respect to plus the imaginary variable.

Differentiating a real function with respect to a complex differential using the differential operator gives the gradient:

$$\begin{bmatrix} \partial x & -\partial y \\ \partial y & \partial x \end{bmatrix} \begin{bmatrix} f & 0 \\ 0 & f \end{bmatrix} = \begin{bmatrix} \frac{\partial f}{\partial x} & -\frac{\partial f}{\partial y} \\ \frac{\partial f}{\partial y} & \frac{\partial f}{\partial x} \end{bmatrix} = grad(f) \quad (10.20)$$

It is established[62] that this kind of differentiation applies within even non-commutative division algebras such as the quaternions. We would expect a form of differentiation which is common to all division algebras. The matrix differential is such a form of differentiation.

The Cauchy-Riemann differential is included:
Clearly, there are some functions like (10.2) which satisfy the Cauchy Riemann equations. These functions are clearly included within the set of matrix differentiable functions, and so Cauchy-Riemann differentiation is included within this more general differentiation. All the established results based on Cauchy-Riemann differentiation still hold true.

---

[62] See : Dennis Morris : Elementary Calculus from an Advanced Standpoint

Differentiation in 2-dimensional space-time:

There are two 2-dimensional division algebras. There are the complex numbers, $\mathbb{C}$, which are often called the Euclidean complex numbers, and there are the hyperbolic complex numbers, $\mathbb{S}$, which are 2-dimensional space-time. Note that 2-dimensional space-time is a division algebra but that our 4-dimensional space-time is not a division algebra:

$$\mathbb{S} = \exp\left(\begin{bmatrix} t & z \\ z & t \end{bmatrix}\right) = \begin{bmatrix} r & 0 \\ 0 & r \end{bmatrix}\begin{bmatrix} \cosh\chi & \sinh\chi \\ \sinh\chi & \cosh\chi \end{bmatrix} \quad (10.21)$$

Differentiation within the hyperbolic complex numbers is properly done in a way similar to that shown above for the complex numbers, (10.11), but, being lazy, we use the hyperbolic differentiation operator:

$$d = \begin{bmatrix} \partial a & \partial b \\ \partial b & \partial a \end{bmatrix} \quad (10.22)$$

We have:

$$\begin{bmatrix} \partial a & \partial b \\ \partial b & \partial a \end{bmatrix}\begin{bmatrix} f(a,b) & g(a,b) \\ g(a,b) & f(a,b) \end{bmatrix} = \begin{bmatrix} \dfrac{\partial f}{\partial a}+\dfrac{\partial g}{\partial b} & \dfrac{\partial g}{\partial a}+\dfrac{\partial f}{\partial b} \\ \dfrac{\partial g}{\partial a}+\dfrac{\partial f}{\partial b} & \dfrac{\partial f}{\partial a}+\dfrac{\partial g}{\partial b} \end{bmatrix} \quad (10.23)$$

These are the 2-dimensional space-time versions of the divergence and the curl.

We have shown the differentiation using the Cartesian form of the hyperbolic complex numbers. We must be wary because the Cartesian form of this algebra is a division algebra only if we impose restrictions of the form $f(a,b) > g(a,b)$. These restrictions are imposed automatically by the trigonometric functions in (10.21).

The Complex Numbers Differentiation Error

Differentiation of a 3-dimensional division algebra:

Consider the commutative 3-dimensional C_3 division algebra[63]:

$$\exp\left(\begin{bmatrix} a & b & c \\ c & a & b \\ b & c & a \end{bmatrix}\right) \quad (10.24)$$

The differentiation operator within this algebra acting upon a potential is (we use the Cartesian form for ease)[64]:

$$\begin{bmatrix} \partial a & \partial b & \partial c \\ \partial c & \partial a & \partial b \\ \partial b & \partial c & \partial a \end{bmatrix} \begin{bmatrix} u(a,b,c) & v(a,b,c) & w(a,b,c) \\ w(a,b,c) & u(a,b,c) & v(a,b,c) \\ v(a,b,c) & w(a,b,c) & u(a,b,c) \end{bmatrix} \quad (10.25)$$

We have:

$$\begin{bmatrix} \partial a & \partial b & \partial c \\ \partial c & \partial a & \partial b \\ \partial b & \partial c & \partial a \end{bmatrix} \begin{bmatrix} u & v & w \\ w & u & v \\ v & w & u \end{bmatrix} \quad (10.26)$$

$$= \begin{bmatrix} \dfrac{\partial u}{\partial a} + \dfrac{\partial w}{\partial b} + \dfrac{\partial v}{\partial c} & \dfrac{\partial v}{\partial a} + \dfrac{\partial u}{\partial b} + \dfrac{\partial w}{\partial c} & \dfrac{\partial w}{\partial a} + \dfrac{\partial v}{\partial b} + \dfrac{\partial u}{\partial c} \\ \dfrac{\partial w}{\partial a} + \dfrac{\partial v}{\partial b} + \dfrac{\partial u}{\partial c} & \dfrac{\partial u}{\partial a} + \dfrac{\partial w}{\partial b} + \dfrac{\partial v}{\partial c} & \dfrac{\partial v}{\partial a} + \dfrac{\partial u}{\partial b} + \dfrac{\partial w}{\partial c} \\ \dfrac{\partial v}{\partial a} + \dfrac{\partial u}{\partial b} + \dfrac{\partial w}{\partial c} & \dfrac{\partial w}{\partial a} + \dfrac{\partial v}{\partial b} + \dfrac{\partial u}{\partial c} & \dfrac{\partial u}{\partial a} + \dfrac{\partial w}{\partial b} + \dfrac{\partial v}{\partial c} \end{bmatrix}$$

(10.27)

$$= \begin{bmatrix} \text{Div} & \text{Curl}_b & \text{Curl}_c \\ \text{Curl}_c & \text{Div} & \text{Curl}_b \\ \text{Curl}_b & \text{Curl}_c & \text{Div} \end{bmatrix}$$

---

[63] See : Dennis Morris : Complex Numbers The Higher Dimensional Forms
[64] The Cartesian form is not a division algebra; only the polar form is the division algebra.

Within this particular spinor space, we see the curl defined in a way which is unusual to our view. This space seemingly plays no part of the physics of our universe.

All differentiation:
The point is that all differentiation can be done with a matrix differentiation operator.

Double differentiation:
It might occur to the reader that we can form a double differentiation operator by squaring the differentiation operator matrix. It does not work.

Summary:
The form of the differential is the form of the product of a variable and the inverse of a variable. As such the differential can be properly defined in only division algebras (spinor algebras) because these are the only algebraic structures which hold 'proper' multiplication.

The form of the differential varies from one division algebra to another division algebra as the product varies from one division algebra to another division algebra.

Since all division algebra multiplication is based on matrix multiplication, the form of the differential will be based on matrix multiplication.

Since all division algebra multiplication is based on matrix multiplication, all differentiation can be done with a matrix differential operator.

The Complex Numbers Differentiation Error

How did mathematicians miss this?:

The first use of the Cauchy-Riemann equations was in 1752 by Jean-Baptiste le Rond d'Alembert[65] (1717-1783) in his study of fluid flow. This was well before Cauchy developed matrix algebra. The same equations were used by Euler in 1797[66] where he used them to define the analytic functions. Cauchy's name became attached to these equations in 1814[67] when he used them to construct his theory of functions. Riemann's name became attached to these equations in 1851 when he used them in his dissertation[68] on the theory of functions.

Within the real numbers, differentiation is associated with a tangent to a function. The Cauchy-Riemann equations are associated with tangents to complex functions within the complex plane. If you want to study fluid flow within the complex plane, you need tangents; this means that you need the Cauchy-Riemann equations.

Tangents to complex functions are also necessary if you are to measure the angle at which two functions cross within the complex plane. This leads to conformal mappings. Without the Cauchy-Riemann equations, we cannot speak of conformal mappings.

It seems that the concept of a differential as a tangent to a function is what drove mathematicians to the Cauchy-Riemann differentiation, and this is a useful concept within the complex numbers. The Cauchy-Riemann equations are not a waste of time.

None-the-less, clearly, the Cauchy-Riemann equations are unsatisfactory in that we can differentiate only a small fraction of the complex functions.

It is hard to understand why Cauchy-Riemann differentiation was not replaced by matrix differentiation 150 years ago or, if not then, 75 years ago. In particular, the use of gauge theory in physics, which dates back

---

[65] d'Alembert J. (1752) Essai d'une nouvelle theorie de la resistance des fluids : Paris
[66] Euler L. (1797) Nova Acta Acad Sci Petrop 10 3-19
[67] Cauchy A. L. (1814) Memoire sur les integrals definies Oeuvres completes Ser 1,1 Paris (published 1882 pp 319-506
[68] Riemann B. (1851) Grundlagen fur eineallgemeine Theorie der Funktionen einer veranderlichen Komplexen Grosse.

to German physicist Hermann Weyl (1885-1955) in 1919[69] cries out for the locally varying complex function to be a potential. Modern $U(1)$ gauge theory assumes that the locally varying $U(1)$ gauge space, which is the complex plane, is a potential. Of course, by the time $U(1)$ gauge theory came along, the Cauchy-Riemann differentiation had been long established and long used in studying fluid flow and conformal mappings.

Summary:

All reasonable functions are differentiable within the complex numbers using the matrix differential. Of these functions, a special sub-set, the analytic functions, satisfy the Cauchy-Riemann equations.

---

[69] Hermann Weyl Annalen der Physik 59, 101 (1919) and Hermann Weyl : Gruppentheorie und Quantenmechanik (1928)

Chapter 11

Non-Commutative Differentiation

In this short chapter, we point out an error of omission which has had consequences with which we will deal later when we look at the Dirac equation.

The differential:

The differential within any algebra is a product, $dy \times \frac{1}{dx}$. We can thus differentiate within only mathematical structures which have a product. Ultimately, the only such structures are division algebras. Some division algebras are multiplicatively commutative, and so, within those commutative algebras, we have:

$$dy \times \frac{1}{dx} = \frac{1}{dx} \times dy \qquad (11.1)$$

Within a multiplicatively non-commutative algebra, we have a non-commutative differential:

$$dy \times \frac{1}{dx} \neq \frac{1}{dx} \times dy \qquad (11.2)$$

Such a simple differential as (11.1) is not well defined within a non-commutative algebra.

Although the quaternions were discovered by William Hamilton in 1843, the differential of a quaternion has been discovered only recently[70]. A non-commutative differential has two fields, named after the electric field and the magnetic field:

---

[70] See : Dennis Morris : The Physics of Empty Space

$$E = \frac{1}{2}(d\Phi + \Phi d)$$
$$B = \frac{1}{2}(d\Phi - \Phi d)$$
(11.3)

wherein Φ is a quaternion potential of the form:

$$\Phi = \begin{bmatrix} \phi & A_x & A_y & A_z \\ -A_x & \phi & -A_z & A_y \\ -A_y & A_z & \phi & -A_x \\ -A_z & -A_y & A_x & \phi \end{bmatrix}$$
(11.4)

and d is the matrix differential operator:

$$d = \begin{bmatrix} \partial t & -\partial x & -\partial y & -\partial z \\ \partial x & \partial t & \partial z & -\partial y \\ \partial y & -\partial z & \partial t & \partial x \\ \partial z & \partial y & -\partial x & \partial t \end{bmatrix}$$
(11.5)

Seemingly, there is evidence[71] that this non-commutative differentiation was discovered over 100 years ago by the mathematician Charles Jasper Joly (1864-1906)[72]. Unfortunately, your author has been unable to verify this. If this is so, then the absence of non-commutative differentiation from the mathematics curricula of universities is an error of forgetfulness. If this is not so, then it is amazing that quaternions can be around for 170 years without anyone thinking of differentiating them.

We will see in due course that the absence of non-commutative differentiation from the consciousness of mathematicians and physicists threw particle physics off course and led to considerable confusion in that area of physics.

---

[71] Michael Jack : Physical space as a quaternion structure, I Maxwell Equations. A Brief Note. arXiv:math-ph/0307038v1 18 Jul 2003
[72] William Rowan Hamilton & C. J. Joly (1899) Elements of Quaternions Vol 1 & 2.

Perhaps we are being too harsh:

If we are going to have two fields as the differential within a non-commutative division algebra, then, since the differential is a product, we must have two 'fields' in the non-commutative product. Only recently has this been understood (unless Joly understood it a 100 years ago). We now understand that the product of two elements, $\{Q_1 \& Q_2\}$ within a non-commutative division algebra (spinor algebra) is of the form:

$$PROD(Q_1, Q_2) = \begin{Bmatrix} \frac{1}{2}(Q_1 Q_2 + Q_2 Q_1) \\ \frac{1}{2}(Q_1 Q_2 - Q_2 Q_1) \end{Bmatrix} = \begin{Bmatrix} E_{Prod} \\ B_{Prod} \end{Bmatrix} \qquad (11.6)$$

Even Mathematicians and Physicists make Mistakes

Chapter 12

The Empty Space Error

The Empty space error is in two parts:

1) The spaces constructed by mathematicians are designated by the number of real axes with complete disregard for the number and type of angles. This dates from a lecture given in 1854 by Riemann in which he presented Riemann geometry to the world. Riemann geometry completely disregards angles. A Riemann space is simply a number of copies of the real numbers.
2) Mathematicians assume they can construct any space they like rather than having to accept what the number one gives them.

Almost universally, mathematicians will tell you that n-dimensional space is formed from n copies of the real numbers. If you ask, "Where do the angles come from?", those same mathematicians will look at you as if you are crazy. Let us consider 3-dimensional space $\mathbb{R}^3$; I'll draw you a picture of it:

There is not an angle in sight. There are no angles in either of the three real number lines. 1-dimensional space has no angles. This is $\mathbb{R}^3$. There is no geometrical structure to this space; as such, we cannot really call it a geometric space.

Angles, and the rotations associated with them, are an important part of the geometry of a space. It is not hard to persuade mathematicians that there are angles in a space. Almost certainly, the mathematicians will immediately assume the angles are 2-dimensional. Without knowledge of the 3-dimensional angles which are within the 3-dimensional division algebras which derive from the C_3 finite group or of any of the other higher dimensional angles, the conventional mathematician has no choice other than to use 2-dimensional angles. There are two types of 2-dimensional angles associated with each of the two types of 2-dimensional division algebras the complex numbers, $\mathbb{C}$, and the hyperbolic complex numbers, $\mathbb{S}$. The mathematician will choose the type of 2-dimensional angles he or she wants.

How many angles will we choose to add to our three copies of the real numbers which we call $\mathbb{R}^3$? Well, if we use 2-dimensional angles, then, to construct a geometric space, we need one angle for every pair of real number lines; that's three in $\mathbb{R}^3$. In $\mathbb{R}^4$, we would need six 2-dimensional angles. Of course, if we were using 3-dimensional angles, the needed number of angles would be the same as the number of trios of axes. We would have only one 3-dimensional angle in $\mathbb{R}^3$ and four 3-dimensional angles in $\mathbb{R}^4$.

There are two types of 2-dimensional angles. Which shall we choose? We are very limited in this choice. The two types of 2-dimensional angles respect the distance functions which are the quadratic forms:

$$d^2 = x^2 + y^2$$
$$d^2 = t^2 - z^2 \tag{12.1}$$

If we are to use 2-dimensional angles, we can have 3-dimensional distance functions of only the forms:

$$d^2 = x^2 + y^2 + z^2$$
$$d^2 = x^2 + y^2 - z^2 \tag{12.2}$$

These are respectively either all signs the same or two signs the same and one sign different. The first of these accepts three Euclidean 2-dimensional angles. The second of these accepts two hyperbolic 2-dimensional angles and one Euclidean 2-dimensional angle. When the mathematician chooses the distance function of the space, he or she also chooses the kind of angles within the space.

Aside: Although one might construct a 3-dimensional space with three Euclidean 2-dimensional angles, one cannot construct a 3-dimensional space with three hyperbolic 2-dimensional angles because we have only two signs (plus or minus).

If the mathematician had chosen the 3-dimensional distance function:

$$d^3 = a^3 + b^3 + c^3 - 3abc \qquad (12.3)$$

then the mathematician would have necessarily have chosen one 3-dimensional angle.

Through the whole of the above, there runs the presumption that mathematicians create mathematics rather than discover mathematics. When a mathematician sets up $\mathbb{R}^3$ space with a chosen distance function, she creates mathematics in the same way that a story-teller creates a fairy story.

Distance functions do not hold angles:
Conventionally, having created $\mathbb{R}^3$ space, the mathematician throws in as many angles as they like as if angles came vacuum packed from the supermarket – just open the fridge and get as many as you want. This is not the correct way to proceed.

Just because a distance function can support a number of angles does not mean the angles come with the distance function. The quaternion emergent expectation space has a 4-dimensional distance function that could support six 2-dimensional angles, but there are only two such angles available, and so we get a space in which rotation is limited to only two of six planes.

In the same way that each separate copy of $\mathbb{R}$ exists, so does each separate angle.

Spinor spaces:
Spinor spaces are the same things as division algebras spaces such as the complex plane.

It is remarkable that mathematicians do not consider the complex plane to be a space even though they have been aware of the complex plane since Wallis presented it in 1673. I'll let you into a little secret; there are no angles in $\mathbb{R}^n$ space; there are only copies of $\mathbb{R}$. Angles exist in only the spinor spaces (division algebras). The same is true of the trigonometric functions associated with those angles and of the rotation matrices which hold those trigonometric functions. There is no rotation in $\mathbb{R}^n$. Rotation matrices exist within a division algebra. Every rotation matrix is an element of a division algebra; for example, the 2-dimensional Euclidean rotation matrix is a complex number:

$$\mathbb{C} = \begin{bmatrix} a & b \\ -b & a \end{bmatrix} \qquad \exp\left(\begin{bmatrix} 0 & \theta \\ -\theta & 0 \end{bmatrix}\right) = \begin{bmatrix} \cos\theta & \sin\theta \\ -\sin\theta & \cos\theta \end{bmatrix}$$
(12.4)

We have seen above how the 3-dimensional rotation matrices of the C_3 finite group are derived analogously to the above 2-dimensional rotation matrix, (12.4). Similar derivations derive all the forms of higher dimensional rotations.

Aside: There are 125 different types of 15-dimensional rotations[73].

Since only spinor spaces have angles, rotation matrices, and trigonometric functions, these are the only spaces with geometric structure. Yet, we do not see 3-dimensional rotations, or 5-dimensional rotations, or rotations of any dimension higher than five within our universe; it is still not clear why we do not see these higher dimensional rotations, but it might be that, like the spinning plate atop a rod we wrote of earlier, physical rotation requires more than one rotation plane. We do see 2-dimensional rotations in our universe. Since 2-dimensional rotations exist in only 2-dimensional spinor spaces, 2-dimensional spinor spaces must be part of our universe.

---

[73] See : Dennis Morris : The Uniqueness of our Space-time.

Our 4-dimensional space-time:

Our observed 4-dimensional space-time is rather strange. It seems to be the only geometric space we observe although perhaps we see the quaternion space within particle physics. Our 4-dimensional space-time is not a spinor space. There is no finite group from which our 4-dimensional space-time derives as a spinor space as is proved by the fact that the distance function is not conserved under multiplication:

$$(t^2 - x^2 - y^2 - z^2)(a^2 - b^2 - c^2 - d^2) \neq T^2 - X^2 - Y^2 - Z^2 \quad (12.5)$$

If our 4-dimensional space-time was a spinor space, it would hold a 4-dimensional angle and the product of two distance functions would be another distance function of the same form.

2-dimensional rotations fit into our 4-dimensional space-time in such a way as to leave two of the dimensions unaffected by the rotation. We have a couple of examples:

$$\begin{bmatrix} \cosh\chi & \sinh\chi & 0 & 0 \\ \sinh\chi & \cosh\chi & 0 & 0 \\ 0 & 0 & 1 & 0 \\ 0 & 0 & 0 & 1 \end{bmatrix}, \begin{bmatrix} 1 & 0 & 0 & 0 \\ 0 & \cos\theta & 0 & \sin\theta \\ 0 & 0 & 1 & 0 \\ 0 & -\sin\theta & 0 & \cos\theta \end{bmatrix} \quad (12.6)$$

Since rotations, angles etc. exist within only spinor spaces, when we rotate our hands within our 4-dimensional space-time, we must be rotating them within a (2-dimensional) spinor space (several spaces actually).

The quaternion emergent expectation space:

We have shown elsewhere[74] that our 4-dimensional space-time is of the same unique structure that emerges from the finite groups as the emergent expectation space of the $C_2 \times C_2$ A_3 algebras. We have also shown that there is a $C_2 \times C_2$ quaternion emergent expectation space.

---

[74] See : Dennis Morris : Upon General Relativity or : Dennis Morris : The Uniqueness of our Space-time.

The Empty Space Error

The A_3 emergent expectation space is the superimposition of six 4-dimensional A_3 algebras. There are thus six angles and four axes in this space – this is our 4-dimensional space-time. We ought to designate it as $\mathbb{R}^4 \theta^3 \chi^3$ to indicate that there are six angles (three of each type).

The quaternion emergent expectation space is the superimposition of only two quaternion algebras. We therefore have a 4-dimensional space which is four copies of the 1-dimensional real numbers held together by only two angles. We ought to designate it as $\mathbb{R}^4 \theta^2$ to indicate that there are two angles (both same type).

Before the quaternion emergent expectation space was discovered, it was tacitly assumed that a 4-dimensional space would have six (2-dimensional) angles. Even after the higher dimensional angles were discovered, it was tacitly assumed that a 4-dimensional space would hold four 3-dimensional angles or one 4-dimensional angle. Only now that we are grasping at an understanding of how the spinor spaces fit together to form our 4-dimensional space-time do we realise that we need angles as well as axes to form a space. There are as many angles as there are isomorphic algebras which together form the emergent expectation space; in the case of the A_3 algebras, this is six; in the case of the quaternion algebras this is two.

Because there are six angles in our 4-dimensional space-time, we can wave our arms around as we please. Since there are only two angles in the quaternion emergent expectation space, rotation in this space is limited to only two Euclidean planes – this, we believe, is why the spin of the electron is directionally quantitised.

The empty space error restated:
If a $\mathbb{R}^n$ space has geometric structure (rotation), then it must incorporate spinor spaces because only spinor spaces have angles. We now have an understanding of how $\mathbb{R}^n$ space emerges from the spinor spaces. We now understand that only two $\mathbb{R}^n$ spaces emerge. These are both 4-dimensional and both emerge from the finite group $C_2 \times C_2$. One

of these spaces emerges from six isomorphic algebras, and so it has six angles; this is our 4-dimensional space-time. The other of these spaces emerges from two isomorphic algebras, and so it has only two angles; this is the quaternion emergent expectation space which seems to hold the weak force and the electron and neutrino.

The empty space error of assuming we can construct $\mathbb{R}^n$ space for any value of n and with total disregard to the number of angles is a consequence of not knowing how $\mathbb{R}^n$ spaces emerge from the spinor spaces. It is entirely reasonable to have made such an error. It is not entirely reasonable to continue making this error.

The Error of Hilbert Spaces

Hilbert spaces were presented to the world by the mathematicians David Hilbert, Frigyes Riesz (1880-1956), and Erhard Schmidt (1886-1959) in the early part of the 20th century. They were later named Hilbert spaces by the mathematician John Von Neumann (1903-1957).

Hilbert spaces are not geometric spaces in that they are not somewhere in which we can wave our arms around. Nor are they geometric spaces in the sense that they are derived from or contain spinor spaces. A Hilbert space is no more than a mathematical concept. A Hilbert space is a way of organising our thinking by associating the objects in our thoughts with positions within an imagined geometric-like space.

Unfortunately, the two fabricated spaces which are our 4-dimensional space-time and the quaternion emergent expectation space are, by the definition of Hilbert spaces, in many ways Hilbert spaces. There is confusion between these two genuine geometric spaces and the concept of Hilbert spaces.

The error of Hilbert space is to:

1) View the two fabricated spaces which emerge from the finite group $C_2 \times C_2$ as Hilbert spaces. These two spaces are the quaternion emergent expectation space and our 4-dimensional space-time.
2) Assume that inner products exist outside of division algebras and the two fabricated spaces. This is to assume that there are inner products where there is no multiplication operation.
3) Assume there are spaces like the two fabricated spaces but of other dimensions and other natures each of which has an inner product.

Hilbert space definition:

A Hilbert space is a vector space with an inner product. A vector space is a set of mathematical objects, functions[75] or vectors[76], together with the operations of addition of vectors and multiplication by a scalar[77] which satisfy the following axioms (have the following properties).

Addition is commutative and associative, and there exist additive inverses (minus vectors) and an additive identity (zero vector). Scalar multiplication is distributive over addition, associative, and there is a multiplicative identity (one).

An inner product of two vectors, $\langle v_1, v_2 \rangle$, is a way of combining two vectors together such that:

$$\langle v_1, v_2 \rangle = \langle v_2, v_1 \rangle^*$$
$$\langle v_1, v_2 \rangle \geq 0 \tag{13.1}$$

The vectors might have complex components; the asterisk denotes conjugation. The inner product must also be linear:

$$\langle av_1 + bv_3, v_3 \rangle = a \langle v_1, v_3 \rangle + b \langle v_2, v_3 \rangle \tag{13.2}$$

The reader might be more accustomed to calling the inner product the dot product.

For example, the set of 2-dimensional vectors in $\mathbb{R}^2$ is a Hilbert space. The inner product on this Hilbert space is the dot product:

$$\begin{bmatrix} a \\ b \end{bmatrix} \cdot \begin{bmatrix} d \\ e \end{bmatrix} = ad + be \in \mathbb{R}^1 \tag{13.3}$$

---

[75] We think of a function as a vector with an infinite number of components – an infinite dimensional vector.
[76] The elements of a Hilbert space are usually called vectors.
[77] A scalar is an element of a division algebra such as the real numbers or the complex numbers.

The Error of Hilbert Spaces

The above, (13.3), is misleading. There is no inner product in $\mathbb{R}^2$ because there are no angles in a copy, or two, of the real numbers. This is really the inner product of two vectors in the complex plane, $\mathbb{C}$:

$$\begin{bmatrix} a & -b \\ b & a \end{bmatrix} \begin{bmatrix} d & e \\ -e & d \end{bmatrix} = \begin{bmatrix} ad+be & ae-bd \\ -(ae-bd) & ad+be \end{bmatrix} \quad (13.4)$$

The above inner product, (13.3), has been copied into the $\mathbb{R}^2$ Hilbert space. This is part of the confusion around Hilbert spaces.

Multiplying vectors together:
We see that there is no multiplication operation defined upon a Hilbert space; we mean there is no way to multiply two vectors together to form another vector. The absence of a multiplication operation is an essential part of the Hilbert space definition. We see that a Hilbert space cannot be a division algebra, and no-one claims it is a division algebra.

The simple fact is that, outside of division algebras, ordered n-tuples of numbers are not a division algebra and so they cannot be multiplied together 'properly'. In 1878, William Kingdon Clifford (1845-1879) developed a way of multiplying vectors together 'properly' and thus formed what are called the Clifford algebras, but the Clifford algebras are really division algebras in disguise[78] and the vectors are really elements of a $C_2 \times C_2 \times ...$ division algebra.

The Hilbert space inner products:
Even though vectors as ordered n-tuples of numbers cannot be multiplied together, mathematicians invented an inner product (dot-product) of ordered n-tuples of numbers by copying the inner products of the 2-dimensional division algebras and analogously extrapolating those inner products into high dimensions. They then added this

---

[78] See : Dennis Morris : The Naked Spinor

invented inner product into the $\mathbb{R}^n$ space associated with the n-tuples of numbers.

Such an inner product is pure invention. Mathematics should be discovered and not invented. Of course, it is right and proper that we invent fairy stories. We are back to the dichotomy between 'natural' mathematics which arises from the number one and 'unnatural' mathematics invented by human beings but which does not really exist.

The Hilbert space error:
The Hilbert space error is one of generalising from the two 4-dimensional emergent expectation spaces which are our 4-dimensional space-time and the quaternion emergent expectation space into all sorts of spaces which do not exist. Underlying the Hilbert space error is the fact that you cannot have an inner product without having a multiplication operation except in the two 4-dimensional emergent expectation spaces which are fabricated from the $C_2 \times C_2$ algebras which are our 4-dimensional space-time and the quaternion emergent expectation space. The nature of both of these spaces is $\mathbb{R}^4$, but the quaternion space can be presented as $\mathbb{C}^2$.

Inner products exist within only division algebras. They are the measure of angle between two vectors and the measure of length of a vector. As an example consider the inner product of two complex numbers (vectors in the complex plane). We have:

$$\langle x+iy, a+ib \rangle = (x-iy)(a+ib)$$

$$\begin{bmatrix} x & -y \\ y & x \end{bmatrix} \begin{bmatrix} a & b \\ -b & a \end{bmatrix} = \begin{bmatrix} ax+by & bx-ay \\ -(bx-ay) & ax+by \end{bmatrix} \qquad (13.5)$$

$$\begin{bmatrix} \cos\phi & -\sin\phi \\ \sin\phi & \cos\phi \end{bmatrix} \begin{bmatrix} \cos\theta & \sin\theta \\ -\sin\theta & \cos\theta \end{bmatrix} = \begin{bmatrix} \cos(\theta-\phi) & \sin(\theta-\phi) \\ -\sin(\theta-\phi) & \cos(\theta-\phi) \end{bmatrix}$$
$$(13.6)$$

$$\cos(\theta-\phi) = ax+by \qquad (13.7)$$

The Error of Hilbert Spaces

We have not worried about normalising the vectors in (13.5).

Since angles exist within only division algebras (spinor spaces), and spinor spaces have a multiplication operation, in general it is in error to form a kind of space without a multiplication operation but with an inner product. The only two exceptions are the fabricated spaces which are our 4-dimensional space-time and the quaternion emergent expectation space.

With a little thought, we see that the way an inner product is formed presumes some kind of multiplication. The inner product is linear, and so it is a linear multiplication – a division algebra multiplication.

In short, the inner product of a Hilbert space is meaningless.

Complex vectors:
There is a second error within the Hilbert space world. This error is the concept of complex vectors. A complex vector is an n-tuple of complex numbers rather like a real vector is an n-tuple of real numbers. This implies that the complex Hilbert space has complex axes. We will look at this in more detail later when we look at Pauli spinors.

Given that the complex numbers, $\mathbb{C}$, are a division algebra like the real numbers, $\mathbb{R}$, then it might seem reasonable to take the view that whatever we can do with the real numbers we can also do with the complex numbers. This view misunderstands the origin of $\mathbb{R}^n$ space.

A vector in $\mathbb{R}^n$ space is an ordered n-tuple of n real numbers. We have seen elsewhere[79] that the only $\mathbb{R}^n$ spaces which emerge from the spinor algebras are the 4-dimensional space-time of our universe and the 4-dimensional quaternion emergent expectation space. Thus, the only vector space is $\mathbb{R}^4$. Coincidently, it is possible to write a quaternion as two complex numbers. It is thus possible, but decidedly questionable, to view the quaternion emergent expectation space as a complex vector space $\mathbb{C}^2$. However, the quaternion emergent expectation space is still

---

[79] See : Dennis Morris : The Uniqueness of our Space-time.

really a $\mathbb{R}^4$ space; the $\mathbb{C}^2$ view is a child of notational obfuscation and confusing coincidence. There is no such thing as a space with complex axes.

Interim summary:
There are only two spaces which do not have a multiplication operation but do have a, fabricated, inner product. These are both 4-dimensional and of a $\mathbb{R}^4$ nature. However, one of these spaces can be viewed as having a $\mathbb{C}^2$ notation.

Hilbert spaces do not have a multiplication operation but do have an invented inner product that is not connected to the non-existent multiplication operation.

How did this error occur?:
Reading Euclid's elements, the reader might be struck with the practical usefulness of the geometry and mathematics. For Euclid, mathematics was firmly set in reality. Mathematics remained close to reality until Cardano discovered the complex numbers, $\mathbb{C}$, in 1545.[80] In the eyes of Cardano and his contemporaries, nothing was counted using complex numbers. Other than particle physicists and electrical engineers, the same view prevails today. So what was Cardano to do with this 'utterly not real' mathematics?

Painting the picture with a far too coarse brush, we might say that the discovery of the complex numbers was when mathematics split from practical reality. But the complex numbers really exist. We cannot simply dump them because they upset our view of reality.

If we are going to entertain the complex numbers, then surely we can entertain non-euclidean spaces in which parallel lines eventually meet or grow further apart. This was done initially without publication by Gauss in 1813 and by Ferdinand Karl Schweikart (1780-1857) in 1818

---

[80] We are far too precise in putting a definite date on the transition of mathematics from reality to the abstract.

and formalised by publication circa 1830 by Janos Bolyai (1802-1860) and Nikolai Ivanovich Lobachevsky (1792-1856).

The idea that parallel lines might meet underlies the idea that empty space can be curved, and this underpins the Riemann geometry which is the mathematics of general relativity.

In 1854, Riemann gave a lecture in which he presented the structure of Riemann geometry. Riemann geometry does not even consider angles. Riemann geometry is not restricted to only 4-dimensional spaces. With the mathematics of Riemann geometry, we can imagine spaces of any dimension. We are now completely divorced from reality. We can fly into a space of any number of dimensions even though such spaces do not exist.

Like children in a sweet shop, mathematicians had been freed from the constraints of reality. Like children in a sweet shop, mathematicians indulged themselves with all they could find. Only now is mathematics starting to look a little queasy.

In a way remarkably parallel to how Riemann geometry freed space from being 4-dimensional, David Hilbert freed Hilbert spaces from the only two fabricated spaces which really exist. A Hilbert space is an abstract concept divorced from reality. Without the constraints of accordance with reality, like children in a sweet shop, we must be very careful not to greedily lead ourselves astray. The Hilbert space error happened because mathematicians became accustomed to looking upon mathematical concepts as existing within their own right irrespective of reality.

Return to reality:
Of course, we now know that the complex numbers discovered by Cardano and initially considered to be separate from reality actually underpin the very realistic 4-dimensional space-time of our universe. We now know that Riemann geometry, in spite of its utter disregard for angles, is the correct mathematics describing our very realistic 4-

dimensional space-time. We just have to restrict Riemann geometry to only our 4-dimensional space-time.

Hilbert spaces, are central to the present formulation of quantum physics. Gladly, we do not need them, and we can reformulate quantum physics using quaternions without even a mention of a Hilbert space. In a sense, the $\mathbb{C}^2$ Hilbert space does exist; it is quaternion space in an obscure and obfuscating notation.

Hilbert spaces are useful:
Like most errors in mathematics, the errors persist because mathematicians have found a use for them. Hilbert spaces are useful in dealing with difficult integrals and in solving differential equations. They also form a crutch to support the mathematical foundations of quantum mechanics.

Summary:
There are only two real spaces which are Hilbert spaces. These are our 4-dimensional space-time and the quaternion emergent expectation space. All the other Hilbert spaces do not exist.

Chapter 14

The Error of Pauli's Spinors

In 1926, Erwin Schrödinger (1887-1961)[81] published the Schrödinger equation[82]. The variable in the Schrödinger equation is a single complex number, $\psi = a + ib$. The Schrödinger equation describes an electron without spin. There is no such thing as an electron without spin. In 1927, Pauli published the Pauli-Schrödinger equation[83]. The Pauli-Schrödinger equation describes an electron with spin. The variable within the Pauli-Schrödinger equation is an ordered pair of complex numbers:

$$\psi = \begin{bmatrix} a + ib \\ c + id \end{bmatrix} \qquad (14.1)$$

The extra complex number is needed to describe the spin. The ordered pair of complex numbers, complex vector if you like, is called a spinor, or a Pauli spinor.

If Pauli had known a little more mathematics, he would have described the spinning electron with a quaternion. An ordered pair of complex numbers is really a quaternion written in obscure notation.

It was not only Pauli who chose to write a quaternion in obscure notation. Dirac and all other physicists of the time did the same. It seems that they simply did not realise that an ordered pair of complex numbers is really a quaternion.

---

[81] A personal note: It is a little discombobulating to realise how many of the great historical figures of mathematics and physics were still alive and teaching many years after my birth.
[82] Schrödinger E. (1926) An Undulatory Theory of the Mechanics of Atoms and Molecules : Physical Review 28 (6) 1049-1070
[83] Wolfgang Pauli (1927) Zur Quantenmechanik des magnetishen Elektrons : Zeitschrift für Physik (43) 601-623

Even Mathematicians and Physicists make Mistakes

Confusion upon confusion:

The reader might recall that we earlier emphasised the importance within mathematics of getting the correct notation. The spinor notation choice of Pauli, Dirac and the other physicists of the time threw a huge obstacle in the way of progress. Of course, it was better that they chose a very bad notation than to have chosen no notation at all. With no notation, there would have been zero progress.

With the spinor notation came the three 2×2 Pauli matrices which are really 4×4 matrices. The three Pauli matrices are separate matrices and are the generators of the Lie group $SU(2)$. The commutation relations of these Pauli matrices are taken to define the Lie group $SU(2)$. The 'Lie group' $SU(2)$ is really just the commutation relations of the quaternions. The 'generators' of the quaternion commutation relations are just the imaginary elements of the quaternions $\{\hat{i}, j, k\}$. We do not need $SU(2)$ or any talk of Lie groups to describe the electron. We have all the commutation relations we need within the quaternions. None-the-less, Pauli's choice of spinor notation led to the inclusion of Lie groups within particle physics; it would be a lot simpler without Lie groups.

The spinor notational confusion was added to by Dirac in 1928[84] when he produced the Dirac spinor as a variable in the Dirac equation:

$$\psi = \begin{bmatrix} \begin{bmatrix} a+ib \\ c+id \end{bmatrix} \\ \begin{bmatrix} e+if \\ g+ih \end{bmatrix} \end{bmatrix} \quad (14.2)$$

The Dirac spinor represents an electron and an anti-electron. It was formed as two ordered pairs to suit the size, 4×4, of the matrices which Dirac used to represent elements of a 16-dimensional Clifford algebra. Dirac completely ignored the fact that the 16-dimensional Clifford

---

[84] Dirac P. A. M. (1928) : The Quantum Theory of the Electron : Proceedings of the Royal Society 117 (778)

The Error of Pauli's Spinors

algebra was, ...er..., 16-dimensional whereas his spinor was only ...er... 8-dimensional.

The unfamiliarity of these 'spinor' things led to confusion that has persisted to the present day. As recently as 1992, the world renowned mathematician Michael Atiyah (1929-) wrote, *"No-one fully understands spinors..."*[85]

As a quaternion:
We have:

$$\begin{bmatrix} a+ib & c+id \\ -c+id & a-ib \end{bmatrix} \equiv \begin{bmatrix} \begin{bmatrix} a & b \\ -b & a \end{bmatrix} & \begin{bmatrix} c & d \\ -d & c \end{bmatrix} \\ \begin{bmatrix} -c & d \\ -d & -c \end{bmatrix} & \begin{bmatrix} a & -b \\ b & a \end{bmatrix} \end{bmatrix} = \begin{bmatrix} a & b & c & d \\ -b & a & -d & c \\ -c & d & a & -b \\ -d & -c & b & a \end{bmatrix}$$

(14.3)

The 4×4 matrix on the right is a left-chiral quaternion.[86]

The message is simple; dump the ordered pair of complex numbers notation and adopt the 4×4 matrix notation of the quaternions. In a later chapter, we will look at the Dirac equation. We will see just how tangled the whole understanding of the electron became because of the Pauli spinor choice of notation. If Pauli had realised that his spinor was really a quaternion, Dirac would have predicted the existence of the neutrino and the violation of parity by the neutrino in 1928.

A notational error:
Pauli did not make a mathematical or physical error when he chose to describe the electron with an ordered pair of complex numbers. The physics was correct. The notation worked just as Roman numerals

---

[85] Quoted in 'The Strangest Man. The Hidden Life of Paul Dirac Quantum Genius' by Graham Farmelo, 2009 pg 430
[86] The details identifying quaternions with Pauli's spinors is given in : Dennis Morris : Quaternions.

work. The error was a practical one; bad notation slowed the advance of particle physics for the next 100 years. Purely notation!!!

To err with notation is human:

We must not be too hard on Pauli. Throughout history, there are many cases of poor notation impeding progress. Nor is it easy to swap notation. We usually need a 'great insight' to drive mathematics and physics forward. The notation usually comes consequent to the 'great insight'. Pauli's great insight would have been the realisation that his spinor was a quaternion.

Another example is the change from Roman numerals to Arabic numerals. The 'great insight' here is the existence of the number base and the various powers of ten. This leads immediately to the existence of the number zero which did not exist in Roman numerals and to the possibility of negative numbers. Following that comes the infinitesimals of calculus.

Another example is the change of notation used for complex numbers. A few years ago, real numbers, complex numbers and quaternions were written as:

$$\mathbb{R}:a \quad \mathbb{C}:a+ib \quad \mathbb{H}:a+ib+jc+kd \qquad (14.4)$$

A change of notation to matrices leads to the real numbers being 1×1 matrices, the complex numbers being 2×2 matrices, and the quaternions being 4×4 matrices. It is now easy to find the 3-dimensional complex numbers.

Chapter 15

The Dirac Equation Error

The Dirac equation was first presented to the world by Paul Dirac in 1928[87]. It is the equation which is taken to describe the electron and the positron and all the other fermions for which parity is a symmetry. The other fermions for which parity is a symmetry are the quarks and their anti-particles and the higher generation versions of the electron which are the muon and the tau particle and their anti-particles.

Prior to the 1950's, it was believed that neutrinos respected parity and were massless, and so it was thought that the Dirac equation described neutrinos when the mass term in the Dirac equation was set to zero. We now know that neutrinos do not respect parity although we can still find (erroneous) presentations of the Dirac equation which set the mass to zero and claim the neutrino is so represented by the Dirac equation.

Aside: It seems that the neutrino field is massless. There is much experimental evidence that neutrinos travel at the speed of light. There is also evidence, the observation of neutrino oscillations, that the squared neutrino field is massive. Since mass is real, any type of complex number with zero real part but non-zero imaginary parts has this nature.[88]

Physicists almost worship the Dirac equation. The Dirac equation was extremely successful in describing the fine details of the hydrogen spectrum, and it seemed to imply the existence of anti-matter which was experimentally confirmed to exist a few years later. The Dirac equation is considered to be very appealing at a theoretical level because it unifies special relativity with quantum mechanics. The Dirac equation is a central part of QED.

---

[87] Dirac P. A. M. (1928) : The Quantum Theory of the Electron : Proceedings of the Royal Society 117 (778)
[88] See : Dennis Morris : The Quaternion Dirac Equation.

But the Dirac equation is wrong:

Unfortunately, the Dirac equation does not make mathematical sense. It is utterly remarkable that an equation that makes no mathematical sense accurately predicts the hydrogen spectrum and other properties of the electron. We will now strip the Dirac equation into its mathematical bits.

The Schrödinger equation:

Erwin Schrödinger took the first step towards the Dirac equation. Schrödinger began with the non-relativistic energy-momentum relation:

$$E = \frac{p^2}{2m} + V \qquad (15.1)$$

This is just the kinetic energy plus an amount of potential energy, V.

Schrödinger then substituted the quantum mechanical differential operators:

$$p = -i\hbar \frac{\partial}{\partial x_i} \qquad E = i\hbar \frac{\partial}{\partial t} \qquad (15.2)$$

into the non-relativistic energy-momentum relation, (15.1), and got:

$$i\hbar \frac{\partial}{\partial t} = -\frac{\hbar^2}{2m} \frac{\partial^2}{\partial x_i^2} + V \qquad (15.3)$$

Giving these differential operators a complex field, $\varphi \in \mathbb{C}$, to operate upon, and setting $\hbar = 1$, gives the Schrödinger equation:

$$i\frac{\partial}{\partial t}\varphi = -\frac{1}{2m}\frac{\partial^2 \varphi}{\partial x_i^2} + V\varphi$$
$$\varphi = a + ib \qquad (15.4)$$

This equation successfully describes a spinless electron. This equation has two degrees of freedom which means that there are only two numbers in the field, the real part and the imaginary part, $\{a,b\}$. There is no such thing as a spinless electron; real electrons have spin.

The Schrödinger equation was later modified by Pauli to describe electrons with spin. He did this by introducing 2×2 'spin matrices' and modifying the field to be a pair of complex numbers presented as a two component complex vector, called a spinor or a Pauli spinor:

$$\varphi = \begin{bmatrix} a+ib \\ c+id \end{bmatrix} \qquad (15.5)$$

This spinor, (15.5), has four degrees of freedom, which means that there are four independent numbers in it, $\{a,b,c,d\}$. Four degrees of freedom is enough to describe an electron with spin-up and an electron with spin-down. Thus, we can describe an electron with spin using a spinor like, (15.5). The modified Schrödinger equation is called the Pauli-Schrödinger equation. Although it was not realised by Pauli at the time, we now know that the ordered pair of complex numbers, (15.5), is a quaternion in obscure notation.

Let us take pause for thought:
There are two 'shocking' aspects to the Schrödinger equation and the Pauli- Schrödinger equation.

The first aspect is that we have an electron field which is a complex number (Schrödinger equation) and a quaternion[89] (Pauli-Schrödinger equation). We might have expected an electron field to be, like the classical electric field, a 4-tuple of real numbers presented as the differential of a vector potential. A classical electric field is a 4-vector, which is four real numbers; an electron field is a quaternion, which is one real number and three imaginary numbers. It is almost as if these are the same things over different types of space.

---

[89] As shown in a previous chapter, a pair of complex numbers is really a quaternion.

The second aspect is that energy and momentum are imaginary differential operators, (15.2). Classically, momentum and energy are:

$$p = mv = m\frac{\partial x_i}{\partial t} \qquad E = \frac{1}{2}mv^2 = \frac{1}{2}m\left(\frac{\partial x_i}{\partial t}\right)^2 \qquad (15.6)$$

This aspect of the theory of quantum mechanics works, but we do not properly understand why the theory works.

Onto relativistic quantum theory:

The Pauli-Schrödinger equation works very well for non-relativistic electrons, but it cannot be the true equation describing the electron because special relativity has effects at all velocities. We need an equation based upon the relativistic energy-momentum relation.

"Great!" said Schrödinger, "let us do the same with the relativistic energy momentum relation as we did with the non-relativistic energy-momentum relation". The relativistic energy-momentum relation is:

$$E^2 = p_x^2 + p_y^2 + p_z^2 + m^2$$
$$m^2 = E^2 - p_x^2 - p_y^2 - p_z^2 \qquad (15.7)$$

This equation effectively says two things.

1) The inner product (dot product) of two (4-dimensional) vectors in quaternion space is $Energy^2$
2) The inner product (dot product) of two (4-dimensional) 4-vectors in our 4-dimensional space-time is $mass^2$

We know that mass and energy are equivalent. The relativistic energy momentum relation, (15.7), suggests that the difference between mass and energy is the difference in underlying space, quaternion space or 4-dimensional space-time.

Since we have (15.7), we also have:

$$E^2\varphi^2 = \left(p_x^2 + p_y^2 + p_z^2 + m^2\right)\varphi^2 \qquad (15.8)$$

wherein φ can be anything. It could be an electron field, for example.

Substituting the quantum mechanical operators, (15.2), directly into this will give the Klein-Gorden equation, but the Klein-Gorden equation cannot represent the electron because, without the spin matrices, it has only two degrees of freedom (the Klein-Gorden field is a single complex number) and we need four degrees of freedom to represent an electron. The Klein-Gorden equation is taken to describe spinless particles like the pi meson. Dirac went another way.

The Dirac equation:
Dirac began with the relativistic energy momentum relation and sought a square root of it to give the energy:

$$E = \sqrt{p_x^2 + p_y^2 + p_z^2 + m^2} \qquad (15.9)$$

So what is the square root of the RHS of the expression, (15.7). Simple, it's a quaternion. Note that we calculate the square of a quaternion, and the square of a complex number of any type, by multiplying the quaternion (complex number) by its conjugate:

$$\text{Quaternion Squared} = QQ^* \qquad (15.10)$$

Although quaternions are non-commutative, a quaternion commutes with its own conjugate. We have:

$$\begin{bmatrix} m & p_x & p_y & p_z \\ -p_x & m & -p_z & p_y \\ -p_y & p_z & m & -p_x \\ -p_z & -p_y & p_x & m \end{bmatrix} \begin{bmatrix} m & -p_x & -p_y & -p_z \\ p_x & m & p_z & -p_y \\ p_y & -p_z & m & p_x \\ p_z & p_y & -p_x & m \end{bmatrix}$$
$$= \begin{bmatrix} m^2 + p_x^2 + p_y^2 + p_z^2 & 0 & 0 & 0 \\ 0 & \sim & 0 & 0 \\ 0 & 0 & \sim & 0 \\ 0 & 0 & 0 & \sim \end{bmatrix} \qquad (15.11)$$

Four roots:

Because a squared quaternion is the quaternion multiplied by its conjugate, we have two roots, the quaternion root and the conjugate quaternion root. These correspond to an electron and an anti-electron (positron).

There are two types of quaternion; these are left-chiral quaternions and right-chiral quaternions, and so we have two pairs of square roots of the relativistic energy momentum relation, (15.7)[90]. We have a pair of left-chiral roots and a pair of right-chiral roots.

The Dirac error:

Unfortunately, Dirac did not see the quaternion solutions. Quaternions were not popular at the time. Quaternions are a 4-dimensional division algebra. Dirac found a 16-dimensional square root. To explain how he did this, we need a small digression into Clifford algebra.

Clifford algebras:

We will construct the 4-dimensional Clifford algebra known as $Cl_{2,0}$. We begin with the expression of a 2-dimensional vector in $\mathbb{R}^2$:

$$x\vec{e_1} + y\vec{e_2} \qquad (15.12)$$

Wherein the two $\vec{e_i}$ are thought of as basis vectors though in actual fact they need not be vectors at all[91]. We square this expression, (15.12), and we insist that the result is a 2-dimensional quadratic form:

$$\left(x\vec{e_1} + y\vec{e_2}\right)\left(x\vec{e_1} + y\vec{e_2}\right) = x^2 + y^2 \qquad (15.13)$$

Aside: When Clifford did this, we was thinking of a right-angled triangle whose hypotenuse, (15.12), was the vector sum of the two vectors representing the other two sides. The expression, (15.12), is a

---

[90] See : Dennis Morris : The Quaternion Dirac Equation.
[91] See : Dennis Morris : The Naked Spinor

vector sum which gives such a hypotenuse. Squaring the hypotenuse vector should give the length of the hypotenuse squared. This is a charming, but rather misleading view; none-the-less, almost by accident, Clifford got some very important algebras which are now named after him.

At the heart of Clifford's expression, (15.13), is the intention to find a way to multiply together two vectors that was 'proper' multiplication meaning it had all the properties like associativity, multiplicative distributivity over addition, absence of zero divisors, multiplicative closure etc. that we associate with algebraic multiplication.

Being careful to keep the order and expanding the LHS of (15.13) gives[92]:

$$x\vec{e_1}x\vec{e_1} + x\vec{e_1}y\vec{e_2} + y\vec{e_2}x\vec{e_1} + y\vec{e_2}y\vec{e_2} = x^2 + y^2 \quad (15.14)$$

The $\{x, y\}$ are just real numbers, and so we can move them around as we like:

$$x^2\vec{e_1}\vec{e_1} + xy(\vec{e_1}\vec{e_2} + \vec{e_2}\vec{e_1}) + y^2\vec{e_2}\vec{e_2} = x^2 + y^2 \quad (15.15)$$

The $\vec{e_1}\vec{e_2}$ terms are called bi-vectors. This, (15.15), implies:

$$\vec{e_1}\vec{e_1} = +1, \qquad \vec{e_2}\vec{e_2} = +1$$
$$\vec{e_1}\vec{e_2} = -\vec{e_2}\vec{e_1} \quad (15.16)$$
$$(\vec{e_1}\vec{e_2})^2 = \vec{e_1}\vec{e_2}\vec{e_1}\vec{e_2} = -\vec{e_1}\vec{e_1}\vec{e_2}\vec{e_2} = -1$$

We have two square roots of unity and one square root of minus unity. This is the 4-dimensional Clifford algebra known as, $Cl_{2,0}$. The subscript indicates that the two basis vectors are square roots of plus unity. The 4-dimensional Clifford algebra, $Cl_{2,0}$ is the set of four basis elements, each of which is multiplied by a real number, $\{a, b, c, d\}$:

---

[92] See : Pertti Lounesto : Clifford Algebras and Spinors

Even Mathematicians and Physicists make Mistakes

$$a.1 + b.\vec{e_1} + c.\vec{e_2} + d.\vec{e_1 e_2} \quad (15.17)$$

This, when properly formulated[93] is a non-commutative A_3 division algebra emerging from the $C_2 \times C_2$ finite group.[94]

If we had chosen the quadratic form to be:

$$(x\vec{e_1} + y\vec{e_2})(x\vec{e_1} + y\vec{e_2}) = -x^2 - y^2 \quad (15.18)$$

we would have derived a quaternion algebra, $Cl_{0,2}$.[95]

Above, (15.12) to (15.18), we have outlined the classic construction of the Clifford algebras. However, this is not the best way to construct the Clifford algebras. Each Clifford algebra is really a division algebra in disguise[96] which is underlain by a finite group of the form $C_2 \times C_2 \times \ldots$. We should really derive the Clifford algebras from the finite groups of the form $C_2 \times C_2 \times \ldots$.[97]

Dirac's first error:
In seeking the square root of (15.7), Dirac formed a 16-dimensional Clifford algebra. He chose the entities $\{\alpha_1, \alpha_2, \alpha_3, \beta\}$ which in conventional Clifford algebra notation would be:

$$\{\alpha_1 = \vec{e_1}, \quad \alpha_2 = \vec{e_2}, \quad \alpha_3 = \vec{e_3}, \quad \beta = \vec{e_4}\} \quad (15.19)$$

He defined the sought for square root to be:

$$Sq.Rt = \alpha_1 p_x + \alpha_2 p_y + \alpha_3 p_z + \beta m \quad (15.20)$$

---

[93] We need to take the exponential to ensure multiplicative inverses and absence of zero-divisors.
[94] See : Dennis Morris : The Naked Spinor
[95] The subscripts in the Clifford algebra names are the numbers of positive signs and the numbers of negative signs in the quadratic distance function. This corresponds to the nature of the basis vectors.
[96] We take the exponential of the classic Clifford algebra to get the division algebra.
[97] This is done in detail in 'The Naked Spinor'.

and he set about discovering what these entities $\{\alpha_1, \alpha_2, \alpha_3, \beta\}$ might be. He squared the square root, (15.20), and set it equal to the required square:

$$(\alpha_1 p_x + \alpha_2 p_y + \alpha_3 p_z + \beta m)^2 = p_x^2 + p_y^2 + p_z^2 + m^2 \quad (15.21)$$

This is a direct analogue of (15.13) and is the classic construction of the 16-dimensional Clifford algebra, $Cl_{4,0}$. It will lead, if properly formulated to a 16-dimensional division algebra from the $C_2 \times C_2 \times C_2 \times C_2$ finite group[98].

Note: There are five classic 16-dimensional Clifford algebras which are based on the quadratic form distance functions:

$$Cl_{4,0} = +t^2 + x^2 + y^2 + z^2 \qquad Cl_{0,4} = -t^2 - x^2 - y^2 - z^2$$
$$Cl_{3,1} = +t^2 + x^2 + y^2 - z^2 \qquad Cl_{1,3} = -t^2 + x^2 + y^2 + z^2$$
$$Cl_{2,2} = +t^2 + x^2 - y^2 - z^2$$
$$(15.22)$$

As division algebras, we have only two non-isomorphic algebras:

$$Cl_{4,0} \cong Cl_{0,4} \cong Cl_{1,3} \cong \{1 + 5\sqrt{+1} + 10\sqrt{-1}\}$$
$$Cl_{3,1} \cong Cl_{2,2} \cong \{1 + 9\sqrt{+1} + 6\sqrt{-1}\} \quad (15.23)$$

It is interesting that the Clifford algebra based upon the quaternion distance function, $Cl_{4,0}$, is algebraically isomorphic to the Clifford algebra based upon the distance function of our 4-dimensional space-time, $Cl_{1,3}$.

Multiplying out the expression (15.21) leads to:

$$\alpha_1^2 = 1, \quad \alpha_2^2 = 1, \quad \alpha_3^2 = 1, \quad \beta^2 = 1$$
$$\alpha_i \alpha_j = -\alpha_j \alpha_i \qquad \alpha_i \beta = -\beta \alpha_i \quad (15.24)$$

---

[98] See : Dennis Morris : The Naked Spinor

This is analogous to (15.16). Clearly, the α_i & β cannot be real numbers. (They are hermitian traceless matrices of even dimension with eigenvalues ± 1.)

Dirac proceeded with[99]:

$$E = \alpha_1 p_x + \alpha_2 p_y + \alpha_3 p_z + \beta m \qquad (15.25)$$

Dirac has a square root. It is a genuine square root, but it is part of a 16-dimensional algebra. Technically, this is a genuine square root only if we formulate the Clifford algebra to be a division algebra, but this is only a small thing of no consequence.

Dirac did not like the mass term having a coefficient which is not a real number; we all know mass is a real number. Multiplying (15.25) throughout by β gives:

$$\beta E = \beta \alpha_1 p_x + \beta \alpha_2 p_y + \beta \alpha_3 p_z + m \qquad (15.26)$$

Technically, the $\beta \alpha_i$ are basis bi-vectors. Dirac changed the notation to:

$$\gamma_0 E = \gamma_1 p_x + \gamma_2 p_y + \gamma_3 p_z + m \qquad (15.27)$$

The gammas are known as the gamma matrices.

Dirac has constructed a 16-dimensional Clifford algebra and written the root of the relativistic energy-momentum relation in this 16-dimensional algebra using one basis vector, $\gamma_0 = \beta \equiv \vec{e_4}$, and three basis bi-vectors, $\gamma_1 = \beta \alpha_1 \equiv \vec{e_4} \vec{e_1}$, $\gamma_2 = \beta \alpha_2 \equiv \vec{e_4} \vec{e_2}$, $\gamma_3 = \beta \alpha_3 \equiv \vec{e_4} \vec{e_3}$. The whole 16-dimensional Clifford algebra is:

---

[99] Dirac. P. A. M. (1982) Principles of Quantum Mechanics. ISBN: 978-0198520115

$$a.1 + b\vec{e_1} + c\vec{e_2} + d\vec{e_3} + e\vec{e_4}$$
$$+ f\vec{e_1}\vec{e_2} + g\vec{e_1}\vec{e_3} + h\vec{e_1}\vec{e_4} + i\vec{e_2}\vec{e_3} + j\vec{e_2}\vec{e_4} + k\vec{e_3}\vec{e_4} \qquad (15.28)$$
$$+ l\vec{e_1}\vec{e_2}\vec{e_3} + m\vec{e_1}\vec{e_2}\vec{e_4} + n\vec{e_1}\vec{e_3}\vec{e_4} + o\vec{e_2}\vec{e_3}\vec{e_4}$$
$$+ p\vec{e_1}\vec{e_2}\vec{e_3}\vec{e_4}$$

Note: It is possible to use an 8-dimensional Clifford algebra to write the root of the relativistic energy momentum equation, but Dirac did not go by this route. Other problems arise if we try to use an 8-dimensional Clifford algebra.

We now have:

$$\gamma_0 E - \gamma_1 P_x - \gamma_2 P_y - \gamma_3 P_z - m = 0 \qquad (15.29)$$

Although the algebra in which this equation, (15.29), is written is 16-dimensional, it can be written using sixteen 4×4 matrices with complex elements, $\in \mathbb{C}$, in which each of the basis elements of the algebra is a 4×4 matrix; for example:

$$\vec{e_4} = \begin{bmatrix} 1 & 0 & 0 & 0 \\ 0 & 1 & 0 & 0 \\ 0 & 0 & -1 & 0 \\ 0 & 0 & 0 & -1 \end{bmatrix}, \quad \vec{e_4}\vec{e_2} = \begin{bmatrix} 0 & 0 & 0 & -i \\ 0 & 0 & i & 0 \\ 0 & i & 0 & 0 \\ -i & 0 & 0 & 0 \end{bmatrix} \qquad (15.30)$$

The 16-dimensional Clifford algebra cannot be written with matrices of a lesser size than 4×4 unless we allow the elements within the matrices to be quaternions. Writing the algebra with 4×4 matrices is just notation; it does not alter the fact that there are sixteen basis elements in this algebra. Above, (15.28), we wrote this 16-dimensional Clifford algebra in the standard Clifford algebra notation.

There is a remarkably common misapprehension within the physics community that the algebra used by Dirac is 4-dimensional because it can be written in sixteen 4×4 matrices. This is an example of how we can look at something that is blindingly obvious and not see it.

Dirac's second error:

The 16-dimensional Clifford algebra formed by Dirac is multiplicatively non-commutative. All Clifford algebras of dimension greater than two are non-commutative. The differential within any algebra is a product, $dy \times \frac{1}{dx}$. Within a multiplicatively non-commutative algebra, we have a non-commutative differential[100]. A non-commutative differential is two fields called respectively an E-field and a B-field.[101] A commutative differential is only one field.

Dirac's second error was to use the commutative differential operators, (15.2) from within the commutative complex numbers, $\mathbb{C}$, rather than the 16-dimensional non-commutative differential operators of the 16-dimensional Clifford algebra. From this point forward. Dirac was going to find only one field; by fudging, he found the electron field; he missed the neutrino field.

Note: The Pauli Schrödinger equation is similarly in error in this regard. Pauli should have used the quaternion non-commutative differential operators.

Inserting the commutative complex number differential operators into the equation, (15.29) gives:

$$i\gamma_0 \partial_t + i\gamma_1 \partial_x + i\gamma_2 \partial_y + i\gamma_3 \partial_z - m = 0 \qquad (15.31)$$

These differential operators need a field of some type, ψ, to operate upon. Dirac found one for them.

---

[100] See: Dennis Morris : Elementary Calculus from an Advanced Standpoint
[101] The E and the B refer to Electric and Magnetic because non−commutative differentiation was first used to deal with classical electromagnetism – see : Dennis Morris : The Physics of Empty Space.

The Dirac Equation Error

Dirac's third error:

Dirac now has an equation written as 4×4 matrices. He needs a four-component object (vector) for the differential operators to act upon. You cannot multiply a 4×4 matrix by a two-component vector.

At this point, Dirac could have chosen a vector with four real components, but everything seems to be complex; there is an i in (15.31), and the Schrödinger equation uses a complex field. The sensible way to proceed is to take a 4-complex-component vector as the field. Such a vector has eight degrees of freedom; that's enough for two electrons with spin. Dirac took the field to be four complex numbers which is a Dirac spinor, also known as a bi-spinor:

$$\psi = \begin{bmatrix} a+ib \\ c+id \\ e+if \\ g+ih \end{bmatrix} \equiv \begin{bmatrix} \begin{bmatrix} a+ib \\ c+id \end{bmatrix} \\ \begin{bmatrix} e+if \\ g+ih \end{bmatrix} \end{bmatrix} \qquad (15.32)$$

Dirac now has two electrons with spin in his field.

The justification for using a four component object, (15.32), is entirely based in the arbitrary use of 4×4 matrices to write the elements of the 16-dimensional Clifford algebra. A 16-dimensional Clifford algebra is more sensibly written as 16×16 matrices, although we could use 8×8 matrices or even 2×2 matrices[102]. The choice of a four component object is often presented as unavoidable given the 4×4 matrices Dirac has chosen – this is a load of nonsense – Dirac should have used 16×16 matrices.

Given Dirac's decision to use 4×4 matrices, and given his decision to use a vector with complex components, $\mathbb{C}^4$, Dirac was bound to find two electrons (two pairs of complex numbers) in his field.

The Dirac equation is:

$$i\gamma_0\partial_t\psi + i\gamma_1\partial_x\psi + i\gamma_2\partial_y\psi + i\gamma_3\partial_z\psi - m\psi = 0 \qquad (15.33)$$

---

[102] We would need to have quaternions as elements of the 2×2 matrices.

The Dirac equation is really four differential equations which couple together the four complex components of the Dirac spinor.

Majorana particles:

The gamma matrices can be written in infinitely many different bases including ones such that every element within the four gamma matrices is imaginary. If the gamma matrices are all imaginary, the Dirac equation is a real equation with real solutions. Within particle physics, if a particle, like the electron, is taken to be a complex number, or a pair of complex numbers, the anti-particle of that particle is taken to be the conjugate of the complex number, or conjugate of the pair of complex numbers; for example, if an electron is a quaternion, then a positron (anti-electron) is a conjugate quaternion. If the solutions of the Dirac equation are real, then an electron is its own anti-particle. Since an electron is not equal to a positron, there is something wrong with the Dirac equation. We cannot avoid this by simply changing the basis in which the gamma matrices are written.

Fermions which are their own anti-fermions are known as Majorana particles after Ettore Majorana (1906-1938). Majorana was the person who first found the four entirely imaginary gamma matrices. One of the confusions within particle physics is the question of whether neutrinos are Majorana neutrinos or not Majorana neutrinos (also known as Dirac neutrinos).

Dirac's great triumph:

It is utterly remarkable that, having got everything wrong, Dirac managed to get the electron field out of this equation. Knowing the answer helped; the electron field is a pair of complex numbers (quaternion). All you have to do is just feed an ordered pair of complex numbers into the equation as the field. Dirac fed two such ordered pairs of complex numbers into the equation as the field.

Dirac's great oversight:

If Dirac had known of non-commutative differentiation, and he had taken the quaternion roots, then, in 1928, he would have predicted:

1) Anti-matter which was not discovered until 1932[103] by Carl David Anderson (1905-1991)
2) The neutrino postulated by Pauli in 1930 and detected by Clyde Cowan (1919-1974) and Frederick Reines (1918-1998) in 1956[104].
3) The violation of parity by the neutrino detected by Chien-Shiung Wu (1912-1997) in 1956[105]

This would indeed have been a remarkable prediction.

Genesis of the error:

How did Dirac miss all this? It might seem that neither Dirac nor any other prominent physicist at the time knew of, or paid much attention to, the quaternions. That is simply not the case. As early as 1929, physicists looked at the Dirac equation and thought, "Aha! quaternions".

What the physicists of the time lacked was an understanding of how one differentiates within a non-commutative algebra.

In the absence of non-commutative differentiate, Dirac used commutative differentiation. This gave him a single field, the electron field, rather than the two fields, electron field and neutrino field.

---

[103] C. D. Anderson : (1933) The Positive Electron Physical Review 43 (6)
[104] C. L. Cowan Jr F. Reines F. B. Harrison H. W. Kruse et al (1956) Detection of the Free Neutrino a Confirmation Science 124 (3212) 103-4
[105] Wu C. S. Ambler E. Hayward R. W. Hoppes D. D. Hudson R. P. (1957) Experimental Test of Parity Conservation in Beta Decay Physical Review 105 (4) 1413-1415

In 1929, Cornelius Lanczos (1893-1974) used pairs of quaternions[106] to rewrite the Dirac equation[107] as the Lanczos equation. Other authors used Clifford algebras to rewrite the Dirac equation[108].[109] In the 1950's, Feza Gürsey (1921-1992) showed the Clifford algebra equations were just degenerate forms of the Lanczos equation[110].[111] In 1956-1958 Gürsey rewrote the Dirac equation with 2×2 matrices with quaternion entries[112], $Mat(2, \mathbb{H})$. There have been numerous other attempts over the decades to rewrite the Dirac equation using quaternions (see Appendix to this chapter). Some of these attempts have been partially successful but often befuddled by complicated and obscure notation. Only recently has the Dirac equation been re-written as a simple quaternion equation.

Aside: Cornelius Lanczos attended the University of Budapest where his physics teacher was Lorand Eötvös[113] (1848-1919). Lanczos graduated in 1915 and received a doctorate in 1921. His major interest was relativity theory and he worked from 1928 to 1929 as Einstein's personal assistant. He remained a personal friend of Einstein for the remainder of his life, and he corresponded regularly with Einstein both as a friend and on matters of relativity. He moved to the USA in 1944 but, consequent to the rise of McCarthyism, he left the USA for Ireland in 1952 where he was appointed head of the theoretical physics department at the Dublin Institute for Advanced Science. He published over 120 papers and books during his lifetime. It is remarkable that this individual is hardly known within the physics community.

---

[106] C. Lanczos. Z. f Phys 57 (1929) 447-483 & 484-493. Reprinted and translated in W. R. Davis et al Cornelius Lanczos Collected papers 2-1132 to 2-1225 arXiv:physics/0508012, arXiv:physics/0508002 arXiv:physics/0508009
[107] A.Gesponer, J.P. Hurni Lanczos equation to replace Dirac's equation pp 509-512 in J. D. Brown (ed) Proceedings of the Cornelius Lanczos Centenary Conference.
[108] F. Sauter Z. F Phys. 63 (1930) 803-814
[109] A Proca J. Phys. Radium 1 (1930) 235-248
[110] F Gürsey PhD Thesis (University of London 1950)
[111] F. Gürsey Phys Rev 77 (1950) 844
[112] F. Gürsey : Correspondence between quaternions and four-spinors. Rev. Fac. Sci. Univ. Istanbul A21 (1956) 33-54
[113] Eötvös is famous for his association with a test of the equivalence of inertial mass and gravitational mass known as the Eötvös experiment.

These attempts to present the Dirac equation as a quaternion equation were frustrated by the lack of a non-commutative form of differentiation which could be applied within the non-commutative quaternions. Attempts to discover the form of non-commutative differentiation began with the work of David Hestenes (1933-) in the 1960's in his development of the geometric calculus, but they were only partially successful. A great step was taken by Peter Jack[114] in 2003, but still the notation is complicated and Jack was not looking directly at the Dirac equation. The recent development of non-commutative differentiation by your author[115] building upon the work of Hestenes and Jack, together with the discovery of the right-chiral quaternions by your author[116], has enabled the rewriting of the Dirac equation as a quaternion equation. This rewrite has brought with it a solution of the neutrino mass problem and an understanding of the violation of parity by the weak force. It might also have brought the first step towards understanding why there are three generations of matter.

Although Dirac set off in the wrong direction when he sought a 16-dimensional root of the relativistic energy momentum equation, this would soon have been corrected by others if the technique of non-commutative differentiation had been available. This error was a case of the physics being ahead of the mathematics.

Most often in the history of maths and physics, the mathematics is ahead of the physics. A clear example of this is Riemann's development of the Riemann geometry which is central to general relativity. In the case of the Dirac equation, the maths was lagging behind.

Appendix – List of papers of interest:
1) Quaternion Formulation of the Dirac Equation   A. S. Rawat, Seema Rawat & O. P. S. Negi   International Journal of Theoretical and Applied Science 3(2) 1-5(2011)

---

[114] Peter Michael Jack : Physical Space as a Quaternion Structure Maxwell Equations, A Brief Note. arXiv:math-ph/0307038v1 18 July 2003
[115] See: Dennis Morris The Physics of Empty Space 2015
[116] See: Dennis Morris Quaternions 2015

2) Quaternion Dirac Equation and Supersymmetry Seema Rawat & O. P. S. Negi arXiv:hep-th/0701131v1 14th January 2007
3) Lanczos equation to replace Dirac's equation? Andre Gsponer & Jean-Pierre Hurni arXiv:hep-ph/0112317v2 18th October 2008
4) Physical Space as a Quaternion Structure Peter M. Jack arXiv:math-ph/0307038v1 18th July 2003
5) Quaternions, Lorentz Group and the Dirac Theory Katsusada Morita arXiv:hep-th/0701074v4 30th March 2007
6) Real Spinor Fields David Hestenes Journal of Mathematical Physics 8 No 4 (1967) 798-808
7) Co-ordinate transformations in quaternion space Zihua Weng arXiv:0809.0126v2 [physics.gen-ph] 15th May 2009
8) Quaternion Quantum Mechanics Arbab I. Arbab arXiv:1003.0075v1 [physics.gen-ph] 27th Feb 2010
9) Quaternionic equation for electromagnetic fields in inhomogeneous media Vladislav V. Kravchenko arXiv:math-ph/0202010v1 6th Feb 2002

Chapter 16

The Scattering Amplitude Error

Within particle physics, one of the major calculations is the probability of something happening. This might be the probability of a particle decaying into other particles or it might be something like the probability of an electron scattering from a positron in a collision.

Electrons scattering from positrons is named after the physicist Homi J. Bhabha (1909-1966) and is called Bhabha scattering.

Part of the calculation of the probability of Bhabha scattering is the expression:

$$M = -e^2 \left(\overline{v_k}\gamma^\mu v_{k'}\right)\frac{1}{(k-k')^2}\left(\overline{u_{p'}}\gamma_\mu u_p\right) + e^2 \left(\overline{v_k}\gamma^\mu u_p\right)\frac{1}{(k+p)^2}\left(\overline{u_{p'}}\gamma_\nu v_{k'}\right)$$

(16.1)

Within this expression, the $\{v,u\}$ terms are Dirac spinors (an ordered 4-tuple of complex numbers) like (15.32). The γ terms are the gamma matrices. The Dirac spinors are 8-dimensional objects. The gamma matrices are elements of a 16-dimensional Clifford algebra. We cannot form products of 16-dimensional objects and 8-dimensional objects. The expression above, (16.1), is mathematical nonsense. The expression above, as far as we can measure, in the practical sense, works perfectly.

There are many other such expressions in the probability calculations of other aspects of particle physics. Each expression is mathematical nonsense and works perfectly.

There is obviously some error here, but we do not know how to correct this error. Until we do know how to correct this error, we bite our tongue and let the physicists get on with it.

We foresee that we will need to rewrite particle physics to be rid of the error in expressions like (16.1), but we do not know how to do the rewrite. It is likely that such a rewrite will solve many outstanding questions within particle physics, and so we are correct to point out the error in such expressions as (16.1). However, we cannot advocate dumping these calculations until we have something to replace them.

A promulgated error:
We have in this case an example of something that is clearly an error, that is known to be an error, but that will none-the-less be promulgated into the foreseeable future.

Origin of this error:
It seems that this error might have been avoided if Pauli and Dirac had used quaternions written in matrix form instead of written as an ordered pair of complex numbers. This would not have been possible without non-commutative differentiation. It is still unknown to your author whether or not the Irish mathematician Joly developed non-commutative differentiation circa 1900, but your author doubts that the matrix notation was used by Joly. As such, it seems this error was unavoidable. The expressions containing this error and the theory, QED, that underpins these expressions has contributed enormously to human understanding over the last century. It is better to err than to do nothing.

Chapter 17

The Lie Group Error

The true nature of the Lie group $SU(2)$ is the quaternion rotation matrix, (7.3). The obscure nature of the Lie group $SU(2)$ is the three Pauli matrices:

$$\sigma_1 = \begin{bmatrix} 0 & 1 \\ 1 & 0 \end{bmatrix}, \quad \sigma_2 = \begin{bmatrix} 0 & -i \\ i & 0 \end{bmatrix}, \quad \sigma_2 = \begin{bmatrix} 1 & 0 \\ 0 & -1 \end{bmatrix} \quad (17.1)$$

Rotation:
Possibly the most important physical law known to humanity is the law that says all physics works the same in all directions in space-time. This is called the isotropy of space-time. The idea is that you can rotate a physical apparatus and it will still work in the same way as it did prior to rotation[117]. Water boils at 100°C regardless of the direction of the kettle's spout. Car engines work the same when the car is heading north as when the car is heading west. The special theory of relativity says simply that physics works the same at all velocities; that is in all directions in 2-dimensional space-time.

Within theoretical physics, this invariance under rotation is stated as 'invariance under the Lorentz transformation'. This is very badly named. The Lorentz transformation of physicists is six separate 2-dimensional rotations; three of these rotations are Euclidean rotations in three orthogonal planes; the other three of these rotations are 2-dimensional space-time rotations in three orthogonal planes. The six rotations are crammed together into a 4×4 matrix and associated with the Lorentz group, $SO(3,1)$. For mathematicians things are different;

---

[117] We ignore gravity.

for mathematicians, a Lorentz transformation is a rotation in 2-dimensional space-time expressed as a 2×2 rotation matrix:

$$\begin{bmatrix} \cosh \chi & \sinh \chi \\ \sinh \chi & \cosh \chi \end{bmatrix} \qquad (17.2)$$

Lorentz invariant physics:

Physicists take the sensible view that any equations describing a physical phenomenon must be Lorentz invariant. If the physics is invariant under rotation, the equations describing the physics should be invariant under rotation.

Great! that makes sense. Our 4-dimensional space-time has only two types of rotation; these are 2-dimensional Euclidean rotation and 2-dimensional space-time rotation as expressed by the matrices:

$$\begin{bmatrix} \cosh \chi & \sinh \chi \\ \sinh \chi & \cosh \chi \end{bmatrix} \qquad \begin{bmatrix} \cos \theta & \sin \theta \\ -\sin \theta & \cos \theta \end{bmatrix} \qquad (17.3)$$

Our 4-dimensional space-time is a fabricated space; this not the only type of space; there are spinor spaces. Associated with our universe, there are four spinor spaces. Two of those spinor spaces are the 2-dimensional complex numbers, $\mathbb{C}$, and the 2-dimensional hyperbolic complex numbers, $\mathbb{S}$, whose rotation matrices are shown immediately above, (17.3). The other two spinor spaces are 4-dimensional; these are the quaternion spaces (two off) and the A_3 spaces (six off).

The essence of gauge theory is that we have to consider rotation in spinor spaces. An angle in a spinor space is called a phase in modern particle physics parlance. Gauge theory says that the forces of nature arise as a consequence of the phase of a spinor space varying from point to point in our 4-dimensional space-time. You can think of this as the axes of the spinor space being rotated at one point in our space-time relative to the axes of the spinor space at a different point in our space-time. There is a conflict between the affine connection (parallel transport) in our 4-dimensional space-time and the affine connection in the spinor space. This conflict of affine connection is a force of nature.

Aside: Special relativity is a global theory in that, in a 4-dimensional space-time without gravity, the positions of two observers in relative motion are irrelevant when we use the Lorentz transformation (rotation in 2-dimensional space-time) to adjust between the physics as observed by the two observers. The two observers can be in any two places in the (without gravity) universe and we get the same result.

The general theory of relativity is a local theory in that the Lorentz transformation we use depends upon the positions within a gravitational field of the two observers. In this sense, the local sense, there is a similarity between general relativity and local gauge theory.

Rotation in a spinor space:

To consider rotation in a spinor space, we need the rotation matrices of the spinor space. The two 4-dimensional rotation matrices are of the forms:

$$\mathbb{H}_{L\chi} = \begin{bmatrix} \cos\lambda & \frac{b}{\lambda}\sin\lambda & \frac{c}{\lambda}\sin\lambda & \frac{d}{\lambda}\sin\lambda \\ -\frac{b}{\lambda}\sin\lambda & \cos\lambda & -\frac{d}{\lambda}\sin\lambda & \frac{c}{\lambda}\sin\lambda \\ -\frac{c}{\lambda}\sin\lambda & \frac{d}{\lambda}\sin\lambda & \cos\lambda & -\frac{b}{\lambda}\sin\lambda \\ -\frac{d}{\lambda}\sin\lambda & -\frac{c}{\lambda}\sin\lambda & \frac{b}{\lambda}\sin\lambda & \cos\lambda \end{bmatrix} \quad (17.4)$$

$$\lambda = \sqrt{b^2 + c^2 + d^2}$$

And:

$$A_3^{L\chi} = \begin{bmatrix} \cosh\lambda & \frac{b}{\lambda}\sinh\lambda & \frac{c}{\lambda}\sinh\lambda & \frac{d}{\lambda}\sinh\lambda \\ -\frac{b}{\lambda}\sinh\lambda & \cosh\lambda & -\frac{d}{\lambda}\sinh\lambda & \frac{c}{\lambda}\sinh\lambda \\ \frac{c}{\lambda}\sinh\lambda & -\frac{d}{\lambda}\sinh\lambda & \cosh\lambda & -\frac{b}{\lambda}\sinh\lambda \\ \frac{d}{\lambda}\sinh\lambda & \frac{c}{\lambda}\sinh\lambda & \frac{b}{\lambda}\sinh\lambda & \cosh\lambda \end{bmatrix}$$

$$\lambda = \sqrt{-b^2 + c^2 + d^2}$$

(17.5)

These are the rotation matrices of the two distinct types of non-commutative algebras of the $C_2 \times C_2$ finite group. This finite group has three order two C_2 sub-groups, and so these algebras have three 2-dimensional sub-algebras.

Spinor rotation as seen from within our space-time:
Within our 4-dimensional space-time, we have only 2-dimensional rotations. So how do the two 4-dimensional rotations above, (17.4) & (17.5), manifest themselves within our 4-dimensional space-time. We do not really know with certainty, but we opine that we have some idea. We opine that the 4-dimensional rotations 'appear to break' into 2-dimensional rotations. Of course, they do not do this within their own space. We assume that the 4-dimensional rotations will 'break' into three 2-dimensional rotations commensurate with the three 2-dimensional sub-algebras.

If the 4-dimensional rotations 'break apart' in our 4-dimensional space-time, we will have one mathematical object associated with each of the broken parts.

The Lie Group Error

Lie groups:

Lie groups are rotational surfaces. An example is the Lie group $U(1)$ which is just the circle. We might view this as the unit circle in the complex plane, $\mathbb{C}$. Another example is the surface of a sphere in the $\mathbb{R}^3$ spatial part of our 4-dimensional space-time; this is the Lie group called $SO(3)$. Another example is the rotational surface within quaternion space. letting $\{b,c,d\}$ range over all values in the quaternion rotation matrix, (17.4), will give this surface; this is called the $SU(2)$ Lie group.

The standard mantra is that there is a set of commutation relations expressed as a Lie bracket associated with each Lie group. The Lie bracket is a way of multiplying together two mathematical objects, $\{x,y\}$. It is:

$$[x,y] = xy - yx \qquad (17.6)$$

The set of commutation relations, together with the objects being multiplied together, is called the Lie algebra of the given Lie group.

Lie groups and infinitesimal transformations:

Lie groups were originally thought of as being generated by infinitesimal rotational transformations. For example, an infinitesimal rotational transformation in the complex plane would be:

$$\begin{bmatrix} \cos\theta & \sin\theta \\ -\sin\theta & \cos\theta \end{bmatrix}\Big|_{\theta\to 0} \to \begin{bmatrix} 1 & 0 \\ 0 & 1 \end{bmatrix} + \begin{bmatrix} 0 & \theta \\ -\theta & 0 \end{bmatrix}\Big|_{\theta\to 0} \qquad (17.7)$$

The unit matrix is ignored and the 'theta matrix' is taken to be an infinitesimal rotational transformation. A finite rotation is taken to be an infinite number of infinitesimal rotational transformations.

Already, we are sinking into a quagmire. An infinite number of infinitesimal rotational transformations can be anything from zero rotation to complete rotation and is rotation through all possible angles at the same time.

To produce a spherical surface in $\mathbb{R}^3$ requires three 'theta matrices' in three orthogonal rotational planes.

Having 'derived' the infinitesimal rotational transformations, a whole area of mathematics called Lie algebra is constructed.

The error of Lie algebras:
Look, we are interested in the rotational surfaces within spinor spaces because physics is invariant under rotation. We can get all the rotational surfaces by taking the rotation matrices of all the spinor algebras that derive from the finite groups. That is all we need. The non-commutative spinor algebras will each have a set of commutation relations. We do not need infinitesimals or Lie algebras.

The use of Lie algebras has led to two large misapprehensions within physics.

The first misapprehension:
The Lie group of $\mathbb{R}^3$ space, $SO(3)$, is three rotation matrices:

$$\begin{bmatrix} \cos\theta & \sin\theta & 0 \\ -\sin\theta & \cos\theta & 0 \\ 0 & 0 & 1 \end{bmatrix}, \begin{bmatrix} \cos\theta & 0 & \sin\theta \\ 0 & 1 & 0 \\ -\sin\theta & 0 & \cos\theta \end{bmatrix}, \begin{bmatrix} 1 & 0 & 0 \\ 0 & \cos\theta & \sin\theta \\ 0 & -\sin\theta & \cos\theta \end{bmatrix}$$

(17.8)

This is three 2-dimensional Euclidean rotations in orthogonal planes.

Looking at the $SU(2)$ quaternion rotation matrix, (17.4), we realise that there are three pseudo-2-dimensional rotations within this matrix corresponding to the three 2-dimensional sub-algebras. These quaternion rotations are double cover rotations in that there are two 2-dimensional rotation matrices within the 4×4 quaternion rotation matrix. For example:

$$\mathbb{H}_{L\chi}|_{c=d=0} = \begin{bmatrix} \cos b & \sin b & 0 & 0 \\ -\sin b & \cos b & 0 & 0 \\ 0 & 0 & \cos b & -\sin b \\ 0 & 0 & \sin b & \cos b \end{bmatrix} \quad (17.9)$$

Thus we seem to have a correspondence between the two Lie groups $SO(3)$ and $SU(2)$. They are both based on a set of three 2-dimensional Euclidean rotations. This correspondence is stated as, "The Lie group $SU(2)$ is a simply connected double cover of the Lie group $SO(3)$". The simply connected means there is only one matrix in $SU(2)$. There is even some mathematics which confirms this relationship between these two Lie groups. The correspondence between these two Lie groups is correct according to the definitions used within the mathematics which proves the correspondence, but in reality it is complete nonsense. We have:

1) $SU(2)$ exists in 4-dimensional space whereas $SO(3)$ exists in 3-dimensional space
2) $SU(2)$ is a 4-dimensional rotation whereas $SO(3)$ is three 2-dimensional rotations
3) There is only one (4-dimensional) angle in $SU(2)$. There are three angles in $SO(3)$

The two rotational surfaces are completely different. The misapprehension is that these two Lie groups have something in common.

The second misapprehension:
All rotation is within spinor algebras (division algebras). If we want an 8-dimensional rotation matrix, we get it from an 8-dimensional division algebra. If we want an 8-dimensional rotation matrix which will 'break' into pseudo 2-dimensional rotations, then we must use the algebras that derive from the $C_2 \times C_2 \times C_2$ finite group because this is the only order eight finite group which has seven order two sub-groups; it also has

seven order four sub-groups, but that is unimportant. The non-commutative algebras that derive from the $C_2 \times C_2 \times C_2$ finite group are all Clifford algebras. All 8-dimensional Clifford algebras have only six non-commutative variables.

Using Lie algebra with its infinitesimals leads to the group $SU(3)$ which has eight non-commutative elements. I'm sorry, but $SU(3)$ does not really exist as a rotational surface.

The whole theoretical construction of quantum electro-dynamics, QED, is built around $SU(2)$ which really exists. The whole theoretical construction of quantum chromo-dynamics, QCD, is built around $SU(3)$. Since $SU(3)$ does not exist, we have a problem. The eight non-commutative elements of $SU(3)$ lead to there being eight gluons. The six non-commutative elements of an 8-dimensional Clifford algebra would lead to six gluons. There are only six things for gluons to do:

$$\begin{array}{ll} red \rightarrow green, & red \rightarrow blue \\ blue \rightarrow green, & blue \rightarrow red \\ green \rightarrow red, & green \rightarrow blue \end{array} \quad (17.10)$$

Perhaps there are only six gluons.

Why do we have Lie groups in physics?:
If Pauli and Dirac has used quaternions in the matrix notation back in the 1920's, we would probably not have conventional Lie groups in physics. Your author opines that physicists chose conventional Lie groups when they ought to have chosen Clifford algebras. Unfortunately, in the absence of non-commutative differentiation, it was not practical at the time to use quaternions or any other Clifford algebra. The physicists were stuck with having to make do as best they were able. Lie groups were accepted in the absence of differentiable Clifford algebras.

We now have non-commutative differentiation. We can now clear away the Lie groups and use the Clifford algebras. Your author opines that this should be done.

Postmethian comments:

Looking over the previous few chapters, we can see that modern particle physics seems to be a tangled and convoluted subject. The tangles are tightened by the fact that modern particle physics gets so much correct; for example, using the scattering amplitude expressions like (16.1) enables physicists to calculate the magnetic moment of the electron accurately to 28 decimal places.

The prospect of undoing such a tangle and still getting the correct numbers from the calculations is a daunting task, yet it must be done if we are not to hit a brick wall and be stopped completely. Perhaps the reader will play a part in undoing this tangle.

Chapter 18

Concluding Remarks

We have discovered that mathematical proof is not quite as airtight and absolute as some mathematicians would have us believe. We have found that even the most simple and clear of mathematical proofs such as the proof of the irrationality of the square root of two given at the start of this book, (1.1), can have unseen surprises. We have seen that some mathematical proofs are so huge and so complicated that there must surely be errors in them somewhere.

We have seen two ways to construct mathematics; these are the axiomatic system and the 'built from the number one' system. Your author prefers the 'built from the number one' system, but your author cannot opine that Euclid's adoption of the axiomatic system was an unproductive error. The axiomatic system has brought immense progress in mathematics and well-being to humankind.

Although the algebraic extension based on a monic minimum polynomial diversion of the previous four centuries is definitely an error, this too is an error which brought progress in mathematics and well-being to humankind.

We have seen that many of the errors that have been made throughout history by mathematicians and physicists are errors which have eventually led to progress and the benefits thereof. We have seen notational errors like Pauli's spinor notation which, technically, are not errors but which have misled mathematicians and physicists. We can blame only ourselves for the continued use of such notation.

We have seen conceptual errors like the 'all rotations are 2-dimensional' error or the all rotations are rotations about an axis error. It is easy to understand why these errors were initially made. It is a tribute to mathematicians that these errors were eventually discovered by mathematics alone; we certainly could not have discovered these errors by observation of our universe.

We have seen undeclared assumptions errors. It is often the realisation that these undeclared assumptions are assumptions rather than absolute truths that is the harbinger of a leap forward in mathematics and physics.

What do we conclude from all we have seen?

Notation:
Changing from using Roman numerals to Arabic numerals led to a blossoming of number theory. Writing finite groups as permutation matrices and, from that writing complex numbers and quaternions as matrices has recently led to a much deeper understanding of rotation and of the nature of empty space. Within only a decade, the change to matrix notation has led to our understanding why our universe is 4-dimensional[118], why it has six 2-dimensional angles, why we have the distance function we do have, and why electron spin is directionally quantitised.

We conclude that notation is very important in mathematics. Good notation simplifies the mathematics and makes everything more lucid. There is a tendency today to use 'concise notation'. I opine that this is an error – I can't understand the damn stuff.

Better to err than to do nothing:
A bad decision is better than no decision. It is better to make an error than to sit around hoping a change of notation or some great insight will waft in through the window.

---

[118] See : Dennis Morris : The Uniqueness of our Space-time.

Don't believe the stark raving obvious:

The stark raving obvious is usually a collation of unrealised assumptions. The difficulty we have is that we do not notice the very obvious aspects of our universe and hence we do not question them. Before Newton, no-one wondered why apples fall to the ground. Before you read this book, you did not wonder why our 4-dimensional space-time contains only 2-dimensional rotations.

We walk a rugged road:

We must accept that we will always be getting things wrong. The path to enlightenment is not easily walked; we have no map to guide us and no lamp to light our way. It is small wonder that we trip and stumble and sometimes inadvertently wander from the road into ditches and marshland. We must not contemn those who have gone before us and have gifted us a piece of bog rather than a metalled road. We would have done what they did in those times. It is ours to pull ourselves from the bog in which we find ourselves.

The biggest error:

In your author's opinion, by far the greatest error of modern mathematics is the continuation of the axiomatic system. We need to adopt the 'built from number one' system started so well by Russell a hundred years ago. This will prune mathematics drastically cutting out all the 'fairy story' maths.

Perhaps the reader will play a part in this historic change of mathematics. It has been a pleasure writing for you.

Dennis Morris

Brotton

August 2016

Other Books by the Same Author

The Naked Spinor – a Rewrite of Clifford Algebra

Spinors exist in Clifford algebras. In this book, we explore the nature of spinors. This book is an excellent introduction to Clifford algebra.

Complex Numbers The Higher Dimensional Forms – Spinor Algebra

In this book, now in its second edition, we explore the higher dimensional forms of complex numbers. These higher dimensional forms are connected very closely to spinors.

Upon General Relativity

In this book, we see how 4-dimensional space-time, gravity, and electromagnetism emerge from the spinor algebras. This is an excellent and easy-paced introduction to general relativity.

From Where Comes the Universe

This is a guide for the lay-person to the physics of empty space.

Empty Space is Amazing Stuff – The Special Theory of Relativity

This book deduces the theory of special relativity from the finite groups. It gives a unique insight into the nature of the 2-dimensional space-time of special relativity.

The Nuts and Bolts of Quantum Mechanics

This is a gentle introduction to quantum mechanics for undergraduates.

Quaternions

This book pulls together the often separate properties of the quaternions. Non-commutative differentiation is covered as is non-commutative rotation and non-commutative inner products along with the quaternion trigonometric functions.

The Uniqueness of our Space-time

This book reports the finding that the only two geometric spaces within the finite groups are the two spaces that together form our universe. This is a startling finding. The nature of geometric space is explained alongside the nature of division algebra space, spinor space. This book is a catalogue of the higher dimensional complex numbers up to dimension fifteen.

Lie Groups and Lie Algebras

This book presents Lie theory from a diametrically different perspective to the usual presentation. This makes the subject much more intuitively obvious and easier to learn. Included is perhaps the clearest and simplest presentation of the true nature of the Lie group $SU(2)$ ever presented.

The Physics of Empty Space

This book presents a comprehensive understanding of empty space. The presence of 2-dimensional rotations in our 4-dimensional space-time is explained. Also included is a very gentle introduction to non-commutative differentiation. Classical electromagetism is deduced from the quaternions.

The Electron

This book presents the quantum field theory view of the electron and the neutrino. This view is radically different from the classical view of the electron presented in most schools and colleges. This book gives a very clear exposition of the Dirac equation including the quaternion rewrite of the Dirac equation. This is an excellent introduction to particle physics for students prior to university, during university and after university courses in physics.

The Quaternion Dirac Equation

This small book (only 40 pages) presents the quaternion form of the Dirac equation. The neutrino mass problem is solved and we gain an explanation of why neutrinos are left-chiral. Much of the material in this book is drawn from 'The Electron'; this material is presented concisely and inexpensively for students already familiar with QFT.

An Essay on the Nature of Space-time

This small and inexpensive volume presents a view of the nature of empty space without the detailed mathematics. The expanding universe and dark energy is discussed.

Elementary Calculus from an Advanced Standpoint

This book rewrites the calculus of the complex numbers in a way that covers all division algebras and makes all continuous complex functions differentiable and integrable. Non-commutative differentiation is covered. Gauge covariant differentiation is covered as is the covariant derivative of general relativity.

Even Mathematicians and Physicists make Mistakes

This book points out what seems to be several important errors of modern physics and modern mathematics. Errors like the misunderstanding of rotation, the failure to teach the higher dimensional complex numbers in most universities, and the mathematical inconsistency of the Dirac equation and some casual errors are discussed. These errors are set in their historical circumstances and there is discussion about why they happened and the consequences of their happening. There is also an interesting chapter on the nature of mathematical proof within our society, and several famous proofs are discussed (without the details).

Finite Groups – A Simple Introduction

This book introduces the reader to finite group theory. Many introductory books on finite groups bury the reader in geometrical examples or in other types of groups and lose the central nature of a finite group. This book sticks firmly with the permutation nature of finite groups and elucidates that nature by the extensive use of permutation matrices. Permutation matrices simplify the subject considerably. This book is probably unique in its use of permutation matrices and therefore unique in its simplicity.

The Left-handed Spinor

This book covers the left-handed parts of mathematics which we call the chiral algebras. These algebras have CP invariance, violation of parity, and many other aspects which makes them relevant to theoretical physics. It is quite a revelation to discover that mathematics is left-handed.

Non-commutative Differentiation and the Commutator

(The Search for the Fermion Content of the Universe)

This book develops the theory of non-commutative differentiation from the fundamentals of algebra. We see what an algebraic operation (addition, multiplication) really is, and we discover that the commutator is a third fundamental algebraic operation within some division algebras. This leads to the first part of the derivation of the fermion content of the universe.

Index

1
15-dimensional rotations, 85
16-dimensional Clifford algebra, 108
1-form, 71

2
2-dim space-time rotation matrix, 42
2-dimensional angles, 83
2-dimensional Lorentz transformation, 44
2-dimensional rotation, 49
2-dimensional rotations, 50
2-dimensional space-time, 38

3
3-dimensional angles, 83
3-dimensional complex numbers, 24, 27, 28, 34, 35, 59
3-dimensional division algebra, 75
3-dimensional numbers, 23
3-dimensional rotation, 50
3-dimensional space, 28
3-dimensional trigonometric functions, 50

4
4-acceleration, norm, 43
4-colour map theorem, 13
4-dimensional Lorentz transformation, 44
4-dimensional space-time, 47, 86
4-dimensional space-time, rotation, 62
4-dimensional space-time, rotations, 86
4-vector, 43

5
5-colour map theorem, 13

8
8-dimensional Clifford algebras, 128

A
A3 algebras, 86
A3 division algebra, 108
A3 emergent expectation space, 87
affine connection, 122
algebraic closure, complex numbers, 24
algebraic extension, 24
algebraic field, 36
algebraic matrix form, 34
Algebraic motors, 36
alleged errors, list of, 5
alternating groups, 9
Ampere, Andre-Marie, 56
amps per metre, unit of magnetic field, 55
analytic functions, 77
Anderson, David, 115
angle, 33
angles, in empty space, 82
angles, popping into existence, 12
Anormal complex numbers, 37
anti-electron, 106
anti-matter, 57, 101
anti-neutrino, 58
anti-particle, 114
Appel, Kenneth, 13
Arabic numerals, 29
Argand diagram, 23
Argand, Jean-Robert, 23

Aristotle, 1
Ars Magna, 23
Aryabhata, 50
Aschbacher, Michael, 9
assumptions, undeclared, 11
Atiyah, Michael, 99
axiomatic system, 18
axioms of a division algebra, 31
axioms, of modern mathematics, 17
axis of rotation, 61

B

BBC, 39
Bhabha scattering, 119
Bhabha, Homi J., 119
Biot, Jean Baptiste, 56
Biot-Savart law, 56
bi-spinor, 113
bi-vectors, 107
Bolyai, Janos, 95
bugs, computer program, 13
building mathematics, 19

C

C2 finite group, 26, 31, 33
C2 X C2 finite group, 27, 64, 87
C3 finite group, 27
Cardano, 23, 94
Cauchy, Augustin-Louis, 1
Cauchy-Riemann differentiation, 73
Cauchy-Riemann equations, 67
Cayley table, 31
Classification Theorem, 9
Clifford algebra, 106
Clifford algebras, 18, 91
Clifford, William Kingdon, 91
Cockle, James, 26, 32, 36
coincidence of distance functions, 47
common understanding, 39, 48
commutation relations, 98, 125
compass needle, 45, 54
compass needle, balance, 45
complex axes, 93
complex function, 71

complex Hilbert space, 93
complex numbers, 33
complex vector, 93
computer proof, 13
conformal mappings, 77
conservation of energy, 45
conventionally integrable, 67
Cowan, Clyde, 115
curl, 70
curl, in space-time, 74
cyclic groups, 35
cyclic groups of prime order, 9

D

d'Alembert, Jean Baptiste le Rond, 77
Davy, Humphrey, 56
De Magnete, by Gilbert, 54
differential, 66, 79
differential of a complex function, 66
differential operator, quaternion, 80
differentiation from first principles, 68
differentiation operator, 72, 74
differentiation operator, 3-dim, 75
differentiation, complex numbers, 69
Dirac equation, 63, 98, 101
Dirac equation, stripped bare, 102
Dirac neutrinos, 114
Dirac spinor, 98, 113, 119
Dirac, Paul Adrien Maurice, 1
divergence, 70
division algebra, 27
division algebras, 33, 76
division algebras spaces, 84
dot product, 90
double cover rotation, 52
double differentiation operator, 76
Dual numbers, 37
Durham university, 40

E

Earth's magnetic field, 45
eigenvectors, 62
Einstein, Albert, 1, 44
electric current, 54

Index

electric field, 57
electromagnetic tensor, 56
electro-magnetism, 43
electron, 46, 52, 63, 97, 106
electron neutrino, 41
electron scattering, 119
electron spin, 63, 87
emergent quaternion space, 47
energy operator, 73
energy-momentum relation, 102
Euclid, 1, 17, 94
Euclidean angles, 33
Euler, Leonard, 14, 77
excluded middle, 3

F

fabricated space, 43
fabricated spaces, 89
fabricated, inner product, 94
Faraday, Michael, 56
Fermat's last theorem, 8
fermions, 114
Feynman, Richard, 1, 15
finite group, 27
finite groups, classification, 9
first principles, complex numbers, 69
fluid flow, 77

G

gamma matrices, 110, 114, 119
gauge theory, 122
Gauss, Johann Karl Friedrich, 1, 6, 24, 94
Gauss, unit of magnetic field, 55
general relativity, 7, 43, 95
generations, of neutrinos, 41
geometric calculus, 117
geometric space, 82
geometric structure, 85
Gilbert, William, 54
gluons, 128
Goldbach conjecture, 14
Goldbach, Christian, 14
good proofs, 15
gradient, 71, 73

Graves, John Thomas, 26
Gürsey, Feza, 116

H

Haken, Wolfgang, 13
Hamilton, William, 26
Heawood, Percy John, 13
Hestenes, David, 117
high end spectrum beta decay, 41
higher dimensional rotations, 51
Hilbert spaces, 89
Hilbert, David, 1, 89
hydrogen spectrum, 101
hyperbolic complex numbers, 26, 32, 74
hyperbolic complex numbers, various
    discoveries, 36

I

imaginary gamma matrice, 114
inertial reference frame, 46
infinitesimal rotational transformations,
    125
inner product, 90
inner product, axioms, 90
intrinsic curvature, 6
irrationality, of root two, 3
isomorphic algebras, 87
isotropy of space-time, 121

J

Jack, Peter, 117
Joly, Charles Jasper, 80, 120
Jordan, Camille, 9

K

Kant, Immanual, 14
Kempe, Alfred, 13
king of theorems, 7
Klein-Gorden equation, 105

L

Lagrange, Joseph-Louis, 25
Lambert, Joann Heinrich, 50
Lanczos equation, 116
Lanczos, Cornelius, 116
Lee, Tsung Dao, 58
left-handed magnetic field, 56
Lie algebra, 125
Lie bracket, 125
Lie group, 98, 125
linear factors, 22
Lobachevsky, Nikolai Ivanovich, 95
local conservation of energy, 46
local gauge theory, 123
London Mathematical Society, 36
Lorentz numbers, 37
Lorentz transformation, 32, 43, 121
Lorentz violating frame experiments, 41
Lounesto, Pertti, 37

M

magnetic dipole, 45
magnetic field, 57
magnetic force, 54
Majorana neutrinos, 114
Majorana, Ettore, 114
mathematics from the number one, 18
matrix differential, 73
matrix differentiation operator, 76
matrix form of complex numbers, 69
minimal polynomials, 22
Minkowski space-time, 43
Minkowski, Hermann, 43
momentum operator, 73
monic minimum polynomial, 22
multiplicative closure, 66
multiplicative closure of matrix form, 35
multiplicative identity, 34
muon, 101
muon neutrino, 41

N

Neumann, John Von, 89

neutrino, 40, 52, 101, 114
Neutrino, 40
neutrino field, 58, 101
neutrino field squared, massive, 42
neutrino mass, 40
neutrino oscillation, 40
neutrino oscillation, formula, 41
neutrino oscillations, 101
Newton, Isaac, 1
non-commutative algebras, 112
non-commutative differential, 112
non-commutative differentiation, 80
non-commutative product, 81
norm, 25
Norman, Robert, 45
north seeking pole, 45
notation, 29, 100

O

octonians, 26

Ɵ

Ɵersted, Hans Christian, 54
Ɵersted, unit of magnetic field, 55

O

operator, differentiation, 72
Opusculorum ad res physicas et mathmaticas pertinentium, 50
ordered pair of complex numbers, 97

P

pair of complex numbers, 63
parallel lines, 94
parity, 56, 101
particle decay, 119
particle physics, 40, 98
Pauli matrices, 98, 121
Pauli spinor, 97
Pauli, Wolfgang, 1, 97, 100
Pauli-Schrödinger equation, 97, 103
permutation matrices, 29, 34

Index

Perplex numbers, 37
Petrus Peregrinus de Maricourt, 54
pi meson, 105
plate spinning, 12
polar form of a division algebra, 49
polynomial equations, 22
poor proofs, 15
positron, 57
potential, complex, 71

Q

QED, 71, 101
quadratic form, 25, 64
quantum mechanical differential operators, 102
quantum mechanics, 104
quarks, 101
quaternion, 97
quaternion algebras, 87
quaternion emergent expectation space, 86
quaternion non-commutative differential operators, 112
quaternion potential, 80
quaternion rotation, 52
quaternion rotation matrix, 51
quaternion, left-chiral, 56
quaternion, right-chiral, 56
quaternions, 26, 27

R

Reines, Frederick, 115
relativistic energy-momentum relation, 104
Riccati, Vincenzo, 50
Riemann geometry, 11, 95
Riemann, Bernhard, 1, 77, 82, 95
Riesz, Frigyes, 89
Romagnosi, Gian Domenico, 55
Roman numerals, 29
rotation, 33
rotation about an axis, 61
rotation matrix, 49
rotation matrix, 3 X 3, 49

rotation matrix, A3, 123
rotation matrix, quaternion, 51, 123
rotation not about an axis, 61
Russell, Bertrand, 1, 19

S

Savart, Felix, 56
scattering, 119
scattering amplitude, 119
Schmidt, Erhard, 89
Schrödinger equation, 63, 97, 102
Schrödinger, Erwin, 1, 97
Schweikart, Ferdinand Karl., 94
simple finite group, 9
simply connected double cover, 127
singular matrices, 34
SO(3), 126, 127
space-time angles, 33
special relativity, 32, 38, 42, 59
special theory of relativity, 121
spin one-half Lorentz transformation, 44
spinning plate, 64
spinning, plate, 12
spinor, 97, 103
spinor algebras, 33, 76
spinor space structure, 60
spinor spaces, 59, 84, 87
split-complex numbers, 37
sporadic groups, 9
square root of 2, 11
standard form Cayley table, 31
Study numbers, 37
SU(2), 98, 121
SU(3), 128
sub-algebra, 28, 34
super-symmetry, 43

T

tau neutrino, 41
tau particle, 101
Tesla, unit of magnetic field, 55
Tessarines, 36
The Letter of Petrus Peregrinus on the Magnet, 54

theorema egregium, 6
theory of functions, 77
trigonometric functions, 33, 85
trigonometric functions, 3-dimensional, 50

U

uncertainty principle, 46
uniform magnetic field, 45

V

vector space, 90
vectors, 91
velocity, 42

violation of parity, 56

W

Wallis, John, 23
wave-particle duality, 46
weak force, 58
Weyl, Hermann, 78
Wiles, Andrew, 7
Wollaston, William Hyde, 56
Wu, Madam Chien-Shiung, 58, 115

Y

Yang, Chen-Ning Franklin, 58

www.ingramcontent.com/pod-product-compliance
Lightning Source LLC
Chambersburg PA
CBHW071817200526
45169CB00018B/347